GEOLOGY IN THE BIBLE

John 1:1-3

By

BILLY R. CALDWELL, Ph.D.

Copyright 2005

Billy Caldwell has asserted his right to be identified as the author of this work under the Copyright, Designs & Patents Act 1988.

All rights reserved. No parts of this publication may be reproduced, stored in a retrieval system, or transmitted in any form or by any means, electronic, mechanical, photocopying, recording or otherwise without the prior permission of the copyright owner.

ISBN 1905363044

First Published in the United Kingdom, May 2005 by Exposure Publishing, an imprint of Meadow Books, 35 Stonefield Way, Burgess Hill, West Sussex, RH15 8DW, UK.

CONTACT INFORMATION

Billy R. Caldwell, Geologist
305 Bodart Lane
Fort Worth, Texas 76108

bcgeology@sbcglobal.net

TABLE OF CONTENTS

ACKNOWLEDGMENTS…………………….. v

CHAPTER ONE……………………………… 7
    CREATION GEOLOGY INTRODUCTION
    (What is Creation Geology?)

CHAPTER TWO……………………………… 12
    GENESIS ACCOUNT OF CREATION
    (The Creation of the Earth)

CHAPTER THREE…………………………… 25
    ROCKS
    (The Study of Rocks and Bible References)

CHAPTER FOUR……………………………. 37
    MINERALS
    (The Study of Mineralogy and Biblical References)

CHAPTER FIVE…………………………….. 44
    FOSSILS
    (The Study of Fossils & Their Relationship to the Bible)

CHAPTER SIX………………………………. 54
    EARTHQUAKES
    (The Study of Earthquakes and Biblical Accounts)

CHAPTER SEVEN………………………….. 70
    SATAN AND CREATION
    (A Theory of Satan at Creation)

CHAPTER EIGHT…………………………… 89
    CONCLUSION
    (A Conclusion Concerning the Matter of Geology and
    Creation With a Creationist Viewpoint)

BIBLIOGRAPHY... **95**

ILLUSTRATION INDEX.............................. **97**

GLOSSARY... **99**

ACKNOWLEDGMENTS

I acknowledge and appreciate the special assistance provided by the following people: Dr. Carl E. Baugh, Wally Clines, Dr. Gene Jeffries, Dr. James Johnson, Andy Mercer, Dr. Jan Mercer, Blanche Miller, Dr. Robert A. Pearle, David Shirley, Michael Shirley and Curtis Tyer. Some have encouraged, some given advice or provided helpful suggestions and some have given much needed computer assistance. Others have read parts of the manuscript.

I would like to pay a very special tribute to my very hardworking, helpful wife, Carolyn. She contributed many long hours on the computer with great devotion to accuracy and corrections. Her help was a labor of love and she worked tirelessly with all the tedious details. I could never thank her enough for her work and encouragement to complete this project.

I extend my special thanks to all of the above who assisted in the review and correcting of portions or all of the manuscript. Any errors of any type are not theirs but my very own.

CHAPTER 1

CREATION GEOLOGY INTRODUCTION

By definition, Geology is a study of the Earth. It is concerned with its origin, history and the dynamics of how it changes. A major question, however, to all the inhabitants on this beautiful Earth, is how did it originate. This creates a great debate in scientific circles that divides the scholars into two philosophies. They are the Creationist and Evolutionist opinions.

A Creationist believes God is sovereign and the Creator of all things in six special 24-hour creation days as recorded in Genesis 1:1-31. God is the only deity that claims to have done this as recorded in Genesis, Chapters 1 and 2. God later reaffirmed this as He gave Moses the Ten Commandments:

> For in six days the Lord made the Heavens and the Earth, the sea and all that is in them, but He rested on the seventh day (Exodus 20:11).

The Darwinian Evolutionist believes the world started with a "big bang" hypothesis or massive explosion in the past; void of God. This theory is very prevalent even though no Evolutionist has ever determined the origin of the mass or the triggering device. This megablast was believed to have occurred 10-15 billion years ago. In most academic universities one is taught to believe that the tiny original mass "somehow" chanced to blow up and heavenly bodies have been revolving, rotating, etc. (with perpetual motion) ever since. The sun, stars and planets are said to have condensed out of gas or dust particles. At some appropriate time with random processes, the raw elements of the Earth are believed to have become living organisms. This evolutionary belief about life is called spontaneous generation. (Note: Spontaneous generation has never been produced in the lab, or ever recorded or observed by anyone and only life produces life.) Evolutionists, however, believe that given enough time this early life (primordial ooze) somehow evolved into single-celled organisms and finally into man through the monkey or ape.

Now, we have listed the two great philosophies of the creation of the Earth and this book will cover both. Everyone must relate to or choose one philosophy or the other. The majority of all the textbooks teach and promote a Godless evolutionary dogma for the Earth's creation. The author stands at this crossroads of life in 2005 with over 50 years of research, seeking the truth as to the origin of the Earth. We all must face this important decision, examine all the evidence and seek truth..."seek and you will find" (Matthew 7:7). Joshua once presented this choice to the people of Israel when he stated in Joshua 24:15, "Choose for yourselves this day whom you will serve." In other words, do you prefer to believe and serve the gods of this world (paganism, secular evolutionary humanism, materialism) or believe and serve the God of creation?

There are really only two basic scientific concepts or philosophies to choose from. One is based on evolution and one on special creation. One, therefore, is humanism, which rejects supernaturalism and embraces man's own humanistic philosophies and the other is theism (belief in God or God centered).

All people must face the principle of Joshua's statement with respect to creation, "Choose for yourselves this day whom you will serve." Joshua chose a side and stated "as for me and my house we will serve the Lord" (Joshua 24:15). The people commented after Joshua's testimony, "We, too, will serve the Lord, because He is our

God." In addition, they said, "We will serve the Lord our God and obey Him" (Joshua 24:18, 24 NIV). Joshua then took a large stone and set it near the Holy Place of the Lord (Joshua 24:26). He stated, "This stone has heard all the words the Lord has said and it will be a witness against you if you are untrue to your God" (Joshua 24:27). This stone was most likely present at the creation of the Earth and observed and witnessed the events. Rocks and stones are, therefore, aware of creation events and even record history to be presented, if necessary. This is verified when Jesus had His triumphal entry into Jerusalem as recorded in Luke 19:29-48. When the Pharisees asked Him to rebuke His disciples for praising Him, Jesus answered, "if they keep quiet the stones will cry out." Some Jews, in the Bible, felt they were safe or righteous because they were descendants of Abraham. Jesus, as the Creator, states He can resurrect Jews themselves from these stones (Matthew 3:9) and for them to not think they can inherit righteousness, but they should repent and produce fruit. This special creation event actually occurred when the first man (Adam) was created out of soil which is weathered and altered rock.

In Habakkuk 2:1-20 the prophet is speaking out against getting rich by evil means, murder and other unrighteousness. They had shamed their names and forfeited their lives. Habakkuk 2:11 states:

> The very stones in the walls of your homes
> cry out against you and the beams in the
> ceilings echo what you say.

All nature knows and obeys God; and it is only man (or mankind) who chooses not to believe. Even though nature displays God's creation, humans are free willed and can choose to not believe or obey. For since earliest times men have seen the Earth and Sky and all God made and have known of His existence and great eternal power. So they will have no excuse when they stand before God on Judgment Day (Romans 1:20).

Many new ideas and "truths" are discovered by carefully planned scientific experiments in laboratories. But, neither of the above philosophies can be proven by a repeatable lab experiment. One, therefore, must just believe or trust in one or the other. Only one is the absolute truth and ones eternal fate is also governed by this choice. The author has chosen to serve, believe and trust God through Christ as my Savior and the Creator of the world. I believe God created all things (John 1:3) and the Bible is truth and can be trusted in all things. It requires faith to believe either philosophy and...

> Without faith it is impossible to please God, because anyone who comes to Him must believe that He exists and that He rewards those who sincerely seek Him (Hebrews 11:6).

By faith we believe that the universe (Earth and stars) was formed at God's command, so that what is seen was not made out of what was visible (Hebrews 11:3).

Therefore, this book is written with a Creationist viewpoint with no apologies. Each subject will be discussed with scientific and Biblical aspects. I believe God is sovereign and the Creator of all things. It is important, therefore, to give credit where credit is due. This research material may improve your knowledge of Geology plus, hopefully, your appreciation of God's omnipotence. It is hoped this knowledge will lead to wisdom. Wisdom, however, is not acquired through study, college degrees, meditation or human experience. "The fear of the Lord, that is wisdom; and to depart from evil is understanding" (Job 28:20). In Him alone "are hid all the treasures of wisdom and knowledge" (Colossians 2:3). God and only the God-head know what happened at the creation of the Earth. Actually, the true history of the Earth is given to us in the Bible. No humans were present at the creation, but God was, and He tells us the events in the Bible. In fact, He did it all Himself, speaking it into existence. Man's humanistic, evolutionary ideas are, therefore, untrue and "professing to be wise they became fools" (Romans 1:22). The Bible also states "The fool hath said in his heart there is no God" (Psalm 14:1). The Apostle Paul warns against those who reject God as living and also as the Creator. He stated, "For the wisdom of this world is foolishness with God" (I Corinthians 3:19) and "I will destroy the wisdom of the wise" (I Corinthians 1:19). Humanistic Earth Science will change continually when new fabrications and discoveries are made and textbooks revised and rewritten. The Biblical account of God as the Creator is unchanging truth..."The grass withers, the flowers fade, but the Word of our God shall stand forever" (Isaiah 40:8).

One can see that the ancient scribe and priest, Ezra, knew the true story of the origin of the Earth and universe when he prayed:

> You alone are the Lord. You made the heavens, even the highest heavens, and all their starry host, the Earth and all that is on it, the seas and all that is in them. You give life to everything, and the multitudes of heaven worship you (Nehemiah 9:6).

Let us, therefore, enter into this *Geology in the Bible* book and explore the splendor of a Holy God-created Earth.

CHAPTER 2

GENESIS ACCOUNT OF CREATION

DISCLAIMER: It is very important to realize that only God and Christ really know exactly what happened (the detailed circumstances) in the creation of the Earth. God's ways are beyond our ways and it is impossible for us to understand His decisions and methods (Romans 11:33). This writer is attempting to theorize or estimate what might have happened from the scientific evidence (especially Geology) we see today. All ideas are presented, therefore, as potentials, suggestions, opinions and theories as to what might have happened in the Geological past using the Bible or God's Word as the true source of all knowledge and wisdom..."In Him lie hidden all the mighty, untapped treasures of wisdom and knowledge" (Colossians 2:3). **END OF DISCLAIMER**

Our study of the Earth must start with its origin and it's location in the solar system. The Genesis account will be our guide and this will be related to current geological data.

The Earth had a special creation beginning (Genesis 1:1) and will have a fiery explosive ending (II Peter 3:11). But then, "According to His promise," God will create "new heavens and a new Earth, wherein dwelleth righteousness" (II Peter 3:13). It was God's desire to create a beautiful world and humankind in His image to enjoy and take care of it. He expected and desired fellowship with mankind. By giving all mankind free will, however, they could accept this gift or reject it. This chapter will deal with the potential events of creation and the entrance and events of sin will follow in a later chapter.

Geology In The Bible

In the original creation story, we read:

> In the beginning God created the Heaven and the Earth. And the Earth was without form and void; and darkness was upon the face of the deep. And the Spirit of God moved upon the face of the waters (Genesis 1:1-2).

In these two magnificent verses we see that God alone created the Heaven and the Earth. He might have created the Earth and most of the solar system, as atomic fiery matter masses. The hot molten mass of the original earth was "without form" until it cooled and it was "void" of anything. The Earth cooled into a sphere or circle shape as stated in the Bible (Isaiah 40:22). As this molten matter cooled, giant steam clouds might have been given off to condense into water and cool the outer crust. This would then form a cooled outer crust (with hot molten lava below that in the mantle) totally surrounded by water and in total darkness. This would be analogous to the description in Genesis 1:1-2. One can drill a well anywhere on the Earth and encounter a layer of granite or basalt at various depths, which originally would have been the hot liquid portion of the Earth, when it was without form and void. Often this layer is on the surface, but it will be encountered by drilling a well to an average of a few thousand feet anywhere on the Earth. During this cooling interval, an erosion and deposition of non-fossiliferous sediments such as granite wash, sandstone, arkose and shale might have occurred.

God is creating, in every case, in a non-time mode. There is no time involved when God is creating, as He is infinite and sovereign. He has, He is and He will always be present. If non-fossil sediments occurred during this cooling period, they might be extensive, but would have been deposited in a short period of time. God can do in one day what might, by natural processes, take a thousand years (II Peter 3:8). Time appears only when God creates and places an item on the Earth, planets, sun or other Earth-related bodies in the universe. It is then placed in a space, time, mass, position that will occur from Genesis 1:3 onward.

A great final portion of this greatest of creation events starts at the end of Genesis 1:2, "And the Spirit of God moved upon the face of the waters." Here in the second verse of the Holy Bible one can see the presence of the third entity of the Trinity (The Holy Spirit). In Genesis 1:26 we see that Christ was present also when God said, "Let Us create man in Our image." Jesus is also known as the Word (I John 5:7) in the New Testament. In John 1:1-3 Christ is described as the Word in creation, "The Word was with God, and the Word was God and all things were made by Him and without Him was not anything made that was made."

God then began creating by speaking things into existence. God called the Universe into existence by His Word. "By the Word of the Lord were the heavens made and all the host of them by the breath of His mouth...For He spoke and it was done; He commanded, and it stood fast" (Psalm 33:6, 9). God's voice itself must have contained the sound, light and elements to produce all things where there was nothing.

No other written revelation to mankind tells how the present time, space and mass universe came into existence. Only the Holy God of the universe was omniscient enough to know what a world or universe should resemble and omnipotent enough to actually create it. This happened in six majestic 24-hour days, not 4.6 billion years. The Hebrew word for day (Yom) usually means a 24-hour day (but doesn't have to mean a literal solar day). But God knew that pagan philosophers would distort that meaning and He added after each creation day, the expression "the evening and the morning," which denotes a 24-hour day. If we let God speak for Himself, He states the six days of creation are symbolic, and since He rested on the seventh day, so should we (Exodus 20:11, Exodus 31:17).

The creation days were as follows:

DAY ONE - "Let there be light and there was light" (Genesis 1:3). God saw the light and it was good. He divided it and called the light day and darkness night. This is not the sun's light since it is to be created on the fourth day. This is light itself...we cannot live without light. It is even a part of Relativity $E=mc^2$ where the "c" is the speed of light. God could not appear on the scene without light. He is the Light of the World. In Him is no darkness at all. If you took an element and purified it completely, it would probably shine. God is so pure, He shines with atomic love. We see this heavenly brilliant nature throughout the Bible. When Moses returned with the Ten Commandments the second time, he had been in the presence of the Lord 40 days and 40 nights. The skin of his face shone brightly from being with God and the people were afraid to come near him. He had to put a veil over his face to overcome their fears (Exodus 34:29-30, 33-35).

At the transfiguration, Christ's glory was again shown when His countenance began to shine and His clothes became dazzling white and blazed with light (Luke 9:29). This was the same Light of the world that later blinded Paul on the road to Damascus (Acts 9:3). It is also recorded that Heaven has no need of the sun or moon to light it, for the glory of God and of the Lamb illuminates it (Revelation 21:23) and there is no night (Revelation 22:5). Light is very mysterious as it can go through cold dark space and, yet, bring light and warmth to planets. It can also dispel darkness. Jesus, being the Light of the World, can also dispel darkness. God revealed light's mysterious nature when He asked Job the following questions. "Where does the light come from, and how do you get there? Or tell me about darkness. Where does it come from? Can you find its boundaries or go to its source?" (Job 38:19-20). Also, "Where is the path to the distribution point of light" (Job 38:24)? Light is so mysterious in nature that it could be a wave or a particle or both.

By stating the evening, then morning was day one there was a beginning of light and darkness or succession of day and night. The Earth was now rotating on its axis and a source of light was present on the Earth, representing the sun, even though the sun was not yet made. The creation of visible light involves the entire electromagnetic spectrum, ultraviolet light, infrared light and other wave phenomena.

DAY TWO - On the second day, God created a firmament (expanse or space) and divided the waters which were under the firmament from the waters above the firmament (Genesis 1:6-8). The Earth was already covered with water as stated in Genesis 1:2. Some of these waters were separated and placed high above the rotating Earth. The upper waters would provide a protective canopy. It could have been a frozen canopy of water or ice which would create a greenhouse effect on the Earth and explain the presence of fossilized ferns in Antarctica and other currently frigid regions. This would produce a perfect climate for all, so that Adam and Eve, even though naked, could be comfortable all year.

A vapor canopy is also a possibility. Both canopies would filter out harmful rays (ultraviolet, cosmic and others) which would contribute to human and animal longevity. This heavy canopy would increase the atmospheric pressure and be effective in preventing disease and promoting good health, such as has been shown in hyperbaric chamber research. Dr. Carl Baugh at the Creation Evidence Museum (www.creationevidence.org) in Glen Rose, Texas is experimenting with this concept and has researched this matter. His work theorizes that the original atmospheric pressure was that of two atmospheres. The two atmospheres of pressure would lead to rapid or instant healing of minor wounds in the early creation period. In every case, God enjoyed His creation and saw that it was good.

DAY THREE - On day three, God said, "Let the waters under the Heaven be gathered together unto one place, and let the dry land appear and it was so" (Genesis 1:9). He called the dry land "Earth" and the waters he called "seas" (Genesis 1:10). Prior to day three, the waters still covered the total Earth. There is now dry land and seas or oceans. These are not the same seas or oceans we have today, since the continents have been rearranged and Noah's flood has changed the surface (II Peter 3:5-6). During this time giant fractures might have occurred that allowed the rocks below the surface to receive the marine or ocean water as the land was uplifted. The original oceans were probably created salty as minerals were dissolved from the original molten Earth. This seawater collected in the porous rocks and most of the rocks today below 1,000 feet contain salt water. During this time, non-fossiliferous rocks were formed as well as a fertile soil (to prepare for plants). The soils were made mature by a combination of erosion and weathering, at beyond the speed of light, and necessary minerals were added for future plant growth. This instant maturity was illustrated when Jesus took new wine and made it instantly old or aged wine (John 2:3-11). We now had light, water and dry land so plant life can be created. On this important third day, we have plants of all types created (Genesis 1:11-13). God created all things fully functional and mature. They automatically had an appearance of age. Fruit trees, herb yielding seed and grass are mentioned. The grass prevented erosion and all plants gave off the much-needed oxygen for future created beings. Plants were the necessary primary food for all the original animals. The plants' seeds were created with all the necessary genes and DNA for reproduction. Each seed was placed in the already created soil and grew in a super fast forward state (maybe the speed of light) to maturity. The trees would have growth rings, but no age until they arrived in the space, mass, time, Earth mode. Even to this day, plants have seeds with programmed genetic codes and reproduce after their own kind, with no indication of evolutionary change. They just keep on rolling along, such as peaches yielding fruit whose seed is in itself and, therefore, when planted, yield peach trees.

DAY FOUR - And God said, "Let there be lights in the firmament of the heaven to divide the day from the night and let them be for signs and for seasons and for days and years" (Genesis 1:14). He made two great lights, the greater light (sun) to rule the day and the lesser light (moon) to rule the night. He made the stars, also (Genesis 1:16).

This created sunlight was now available for energy and plant life. A source of light for plants was also available on the third creation day. We now have gone from "let there be light", the creation of light to "let there be lights" (light-givers). Even though plants were created on the third day, they only had one day to wait for the official sun source of light. Since all the lights including the stars were for signs and seasons, the light-trails from the stars were created at the same time. This made the stars visible instantly, stars that are now many light years away. This seems incredible but "Is anything too hard for God" (Genesis 18:14). The answer is "No" (see Luke 1:37 and Jeremiah 32:17 and 27). The wise men from the east might have been told by Daniel the Prophet, that a certain star would be the sign of the Savior (Daniel 2:48, Daniel 5:11, Daniel 7:13-14 and

Matthew 2:1-2). Wise men still seek Him today. By faith we believe the creation account to be true. It is important to mention that two lights were created. The sun generates it's own light and the moon reflects light, so both give light as stated in Genesis 1:16.

DAY FIVE - And God said, "Let the waters bring forth abundantly the moving creature that hath life and fowl that may fly above the Earth in the open firmament of Heaven. And God created great whales (and great sea monsters) and every winged fowl after his kind" (Genesis 1:20-21). Animal (marine and terrestrial) life now had abundant food, like grass, fruit, seaweed and algae, to grow and reproduce "after their own kind"; which they have done unto this day. The first verbal command was given here by God, to these created beings (water animals, birds) to be fruitful and multiply (what He will also tell mankind later) and to fill the waters and multiply in the Earth (Genesis 1:22). The Earth was now ready to be covered throughout with plants to supply food for land animals and mankind, which will follow. Fish would also be a later source of food and they needed to be abundant throughout the world. The great sea monsters Plesiosaur, Mosasaurus, Elasmosaurus and the Leviathan were created during this time. These animals lived in the water and were possibly described in Job 41 where all 34 verses describe a fearless beast of the seas. The smelly monster caught off the coast of New Zealand by a Japanese fishing boat in 1977, might have been one of the sea monsters which probably still exists today (Dennis R. Peterson, *Unlocking the Mysteries of Creation*, p. 146). The Loch Ness Monster might also be in this category (William J. Gibbons, Ph.D. and Dr. Kent Hovind, *Claws, Jaws and Dinosaurs*, pp. 9-17).

DAY SIX - And God said, "Let the Earth bring forth the living creature after his kind, cattle and creeping things, and beast of the Earth after his kind and it was so" (Gen. 1:24).

God had now created green plants (Day Three), sunlight (Day Four) and could now create a great variety of plant eating land animals. We can see His intelligent design in all the variety of land animals and plants on this Earth and the logic of His orderly arrangement. He created innate, inborn traits or instincts (Job 38:36), so animals would instantly know how to function, even those who had never seen their parents...examples are sea turtles, alligators and salmon. Once God had created the special organs He liked, such as an eye or an ear, He placed them on most animals. There was no evolutionary struggle for improvement concerning these animals. They could not change to different kinds because God made each group "after its kind." The DNA module was programmed to allow genetic variation, so each kind could have individual variation, but not beyond the kind itself. For example, from the St. Bernard to the Chihuahua, they are all varieties of dogs or dog kind. Animals still reproduce after their own kind. Evolutionary crossover from one genus to another has never occurred, past or present. Insects were most likely created at this time to pollinate the ubiquitous variety of plants that would be the original food for all created animals, including mankind (Genesis 1:29-30).

The world was now ready for human inhabitants. And God said, "Let us make man in our image" (Genesis 1:26). The Trinity was in counsel about the creation of man in Its image. Man was, therefore, complex and highly organized, but of a higher order, since he was given a spirit as well as a soul. Man was to have dominion over every living thing on the Earth. Mankind was to be fruitful and multiply and replenish the Earth and subdue it (Genesis 1:28). There was no sin, sorrow, pain, suffering or disease at this time. Plants were given for food for all animals including man (Genesis 1:29-30). An abundant variety of plants were created for food on the Third Day. Man was not created millions of years ago by evolving from monkeys, but from dust (the elements of the Earth) by the creative hand of God. God breathed into him the breath or spirit of life and man became a living being (Genesis 2:7, I Corinthians 15:45-49). No missing link fossil has ever been found. A very limited amount of chimp or ape fossils have been unearthed, which have been falsely classified as missing links between man and the monkey. The Neanderthal Man and Cro-Magnon Man fossils have been found to be similar to modern Europeans (Duane Gish, Ph.D., *Evolution: The Fossils Still Say NO!,* p. 305).

God was pleased with all His work.

> And God saw everything that He had made, and, behold it was very good. And the evening and the morning were the Sixth Day (Genesis 1:31).

When all the work was done in six creation days He rested:

> And on the Seventh Day God ended His work which He had made; and He rested on the Seventh Day from all His work which He had made (Genesis 2:2).

God blessed the Seventh Day and sanctified it and told us to observe the Sabbath as a Holy Day (Exodus 20:11) and included this as one of His Ten Commandments.

The perfect world of creation where there was no pain, sorrow, disease, suffering, decay or death no longer exists. God created a perfect world, but man disobeyed God and sin then entered the world. This allowed the entrance of pollution, contamination, death and decay which exists today. The second law of thermodynamics, which states that all processes tend toward a state of decay and ultimate death, now went into effect. The curse of God changed all of nature and allowed thorns, thistles and death. Plants and animals changed and some animals became carnivorous. Diseases were later created by sinful behavior. Nature has been suffering every since from sin and waits patiently for freedom and change (Romans 8:18-22). Man has polluted the world so much it must be destroyed and a new heaven and a new Earth created (II Peter 3:10-13).

The fall of man enabled sin to enter the world and allowed Satan to be the leader of that domain. God permitted Satan to exist and lead this rebellion on the Earth, for a season. Mankind, as free will creations, could now become followers or allies to Satan or to God. All future beings would be born into a sin-polluted world. Satan would deceive and lie his way into popularity and become the Prince of this world (John 14:30). Hell would be created for Satan and other fallen angels (Matthew 25:41) and Heaven for the followers of God. No one is sent to either place. Each individual will go to their chosen group's destiny. The Devil (Satan) and his followers will eventually be cast into the Lake of Fire to be tormented day and night forever (Revelation 20:10). Those who accept the Son of God (Christ) as their personal Savior will have their names written in the Lamb's Book of Life (Revelation 20:12, Revelation 21:27). God will personally welcome these believers to a beautiful Heaven (Revelation 21:3) to be with Him throughout eternity. In Heaven there will be no more curse (Revelation 22:3), pain or sorrow (Revelation 21:4). The greatest and most important decision of anyone's life is to choose correctly your eternal destiny.

It is not known if the Earth is the center of the universe; but we do know the Earth is the center of interest in the universe, to God. Because, after the fall of mankind He sent His only Son as the Savior of the World.

> For God so loved the World that He gave His only begotten Son, that whosoever believeth in Him should not perish but have everlasting life (John 3:16).

The Geology of the past was covered by the flood deposits of Noah's time which will be discussed in future chapters. We will never determine the exact or total nature and origin of all of God's creation. Three thousand years ago the Psalmist was seeking to understand the past and present when he stated, "I have considered the days of old, the years of ancient times" (Psalm 77:5). Everyone needs to spend less time worrying and wondering about origins and start working and worshipping. Christians will have the joy of seeing and understanding everything in eternity (all by and by). Only then will the believers see and hear the true story of the creation of the Earth, from the Great Creator Himself. We are to enjoy and marvel at the Earth's wonderful splendor and variations showing unimaginative creative design. The present beauty of our created Earth is a hint of how wonderful the good Earth was before sin entered and the great curse occurred (Genesis 3:17-18). The beautiful fall foliage, spring flowers and sunsets should only strengthen our belief in an almighty, omnipotent, creative and loving God who chose to make a beautiful world for us to share. Nature should turn us to God instead of away from Him. Everyone should see the Earth's splendor, stand in awe and then speak as the Psalmist David when he wrote,

> When I consider thy heavens, the work of thy fingers, the moon and the stars, which thou hast ordained; What is man, that thou art mindful of Him? And the son of man, that thou visitest him? (Psalm 8:3-4).

Scientists work with "how", "when" and "where"; but they cannot answer the question of "why." "Why" can only be totally answered by God. And who could be His counselor or teach Him knowledge and understanding in creation (Isaiah 40:12-14). The answer is, it is impossible for a finite man to understand the decisions of an infinite God. His ways are past finding out (Romans 11:33). God seemingly created a beautiful world to have fellowship with mankind and for us to see His glory. Geological and scientific discoveries only reveal the inventive, creative, nature of God. The world's marvelous design should only bring glory to God. The marvels of nature "should" draw men closer to God instead of leading mankind away from Him. "Oh, what a wonderful God we have! How great are His wisdom and knowledge and riches! For everything comes from God alone. Everything lives by His power, and everything is for His glory" (Romans 11:33-36).

CHAPTER 3

ROCKS

After years of studying and examining rocks throughout the world (and now the moon), scientists have been able to come up with only three principle categories. These categories of rocks are divided into groups based on the major Earth processes which formed them. The *Dictionary of Geological Terms*, prepared by the American Geological Institute, defines a rock as "strictly any naturally formed aggregate or mass of mineral matter, whether or not coherent, constituting an essential and appreciable part of the Earth's crust." Rocks, therefore, are the very substance of the Earth. They are composed of the same elementary particles as all other matter in the universe. The masses of rock, however, are very extensive and cover the whole Earth. They may be exposed on the surface as a layer or strata projecting through the detritus and soil (an outcrop), be below the soil or covered by the ocean. The first layer of solid rock below the soil is called bedrock.

Since recorded history is available for only about 6,000-7,000 years (apart from the Bible), the Geologist must read the rock history for the rest of the story. Rocks differ greatly from place to place because of the many rock-forming processes. Geologists seek to know the composition and distribution of rocks and how they are formed, destroyed and how they are uplifted into continents and depressed into ocean basins. Geologists have studied the rocks for many years and have pieced together a geological history of the Earth from this information. This study has been totally influenced by the humanistic ideas that leave God's Word and His existence out of the equation. The Bible is not considered, since that is deemed as religion and humanistic scientists think they do not mix. Uniformitarianism states that past geological events are to be explained by those same physical principles that are observed today. In other words, "the present is the key to the past." The concept and belief in uniformitarianism has also led geological history astray. The Bible, however, records a different view of creation defined as catastrophism.

The doctrine of catastrophism states that sudden, violent, short-lived, more or less worldwide events, outside our present experience of the knowledge of nature, have greatly modified the Earth's crust. Noah's flood, the greatest Biblically-recorded catastrophe, would be accepted by the catastrophist and rejected by the uniformitarianist. Each scientific idea or theory is strongly biased or influenced by whether the God of Creation is considered as being involved. As a catastrophist, the author believes some of the great granite masses, which were later uplifted into dry land and mountains, could have formed and cooled between Genesis 1:1-2. This is not the same as the Gap Theory, which states that millions of years were involved. This would, instead, be instant maturity like Jesus changing new wine to old wine (John 2:8-10).

D. Russell Humphreys, Ph.D., in his current research called the RATE project, finds evidence for a young world. This work, Impact Art. #352, was published for the Institute for Creation Research in October 2002. The results of this work strongly support an accelerated decay hypothesis, that episodes with billion-fold speed-ups of nuclear decay have occurred in the recent past, such as the early Creation week. He writes, "Such accelerations would shrink the alleged 4.5 billion year radioisotope age of the earth down to the 6,000 years that a straightforward reading of the Bible gives." The flood of Noah's time would also form great layers of sediments in a year, instead of millions of years.

As stated earlier, Geologists classify rocks in three

types or categories according to the major Earth processes which form them. The three types are Igneous, Sedimentary and Metamorphic rocks. Each rock type will be covered separately.

IGNEOUS ROCKS - Rocks formed by the cooling of original hot, molten material. These rocks are called igneous rocks from the Latin word *ignis* meaning fire. The Earth's crust cooled but molten rocks are still below the surface everywhere, it's just a matter of depth. The Bible verifies this by stating that men know how to obtain food from the surface of the Earth while underneath there is fire (Job 28:5). All the planets and the moon seem to have had a fiery igneous rock or molten origin. The Earth also was once most likely molten and had no shape and was void (Genesis 1:2). This explains the abundance of massive volumes of igneous rocks throughout the world, below the sedimentary rocks. These huge granite blocks uplifted and formed the central mass of many mountain ranges. The very roots of the continents consist of rock which cooled from a liquid magma state and crystallized into igneous rocks, especially granite.

Magma, a fiery glowing fluid, flows from the subsurface and forms volcanoes when it reaches the surface. This magma is mostly bodies of silica melt or silicates (minerals with crystal structure containing SiO_4). The igneous rocks, formed from cooled magma that escaped to the surface as lava, are called extrusive volcanic rocks. Such rocks are usually composed of very tiny crystals since they cool rapidly. Some lava erupts into water and cools so quickly that crystallization cannot occur at all. This results in a glassy rock such as obsidian. Because the magma is less dense than the surrounding rocks, it tends to rise within the crust. If it is under pressure and charged with gases, violent eruptions can occur. The magmas that erupt in volcanoes result in the forming of rocks, such as scoria, basalt and rhyolite. Most of the Earth's oceanic crust is formed of basalt. One type of igneous rock is full of unconnected cavities, formed during cooling, and it is called pumice. Most chunks of pumice will float on water. If the lava or volcanic eruption flows quiescently as a blocky front, it will form chunky pieces called *aa* type lava. If the lava congeals into ropy structures or wavy masses, it is referred to as *pahoehoe* type lava.

Most of the islands in the Atlantic and Pacific Oceans are the result of volcanic action, where eruption deposits continued above the ocean level. Some of these islands still contain active volcanoes. Magma or lava flows are most active at weak spots in the crust and along plate tectonic boundaries. The ring of fire around the Pacific Ocean is an example. When magma or molten rock cools slowly, usually at depths below 1,000 feet, crystals separate from the molten liquid and coarse-grained rocks such as granite occur. Generally speaking, the slower the cooling the larger the crystals. The magma masses that harden below the surface are called intrusive rocks or bodies. Due to the Earth being hotter at depth, magma is available in the mantle of the Earth, for volcanic eruptions at any time.

Rocks brought back from the moon by the Apollo 11 and Apollo 12 missions were igneous in nature. They included two main igneous rock types, a vesicular basalt and a fine grained gabbro.

SEDIMENTARY ROCKS - Sedimentary Rocks are classified as clastic (consisting of fragments of pre-existing rocks) or nonclastic. The main clastic rocks (sandstone and shale) can be composed of rock fragments of any type of previous rock. They are formed at the surface of the Earth, either in water or on the land. They are layered accumulations of sediments, very fine to coarse fragments of rock, minerals, precipitated chemical matter or plant and animal material. Examples of clastic sedimentary rocks are clay, silt, sandstone and gravel. The nonclastic types are formed by chemical precipitation, by biological precipitation and by the accumulation of organic material. These processes extract specific materials from their surroundings, usually seawater, and precipitate these substances, forming rocks. At no time during their formation are temperatures or pressures especially high and their mineral constitution and physical appearance reflect this fact. These rocks are normally deposited parallel or nearly parallel to the Earth's surface. If they are at high angles to the surface or twisted or broken, then some kind of Earth movement has occurred since their deposition. Examples of nonclastic sedimentary rocks are limestone, dolomite, chert, gypsum, rock salt and coal.

One of the most interesting things about sedimentary rocks is they contain most of the world's fossils. These fossils can tell us much about the depositional environment. The story of fossils, however, will be contained in a separate chapter.

METAMORPHIC ROCKS - Metamorphic rocks are rocks that have been changed while in the solid state, either in texture or in mineral composition. In simple terms, a metamorphic rock is a rock that has been changed by heat and/or pressure. Metamorphic is from the Greek word *meta* (which means change) and *morphe* (which means to change form). Any type of original rock can be changed into a metamorphic rock, and there are two basic types of metamorphism: Contact and Regional. Contact metamorphism occurs at or near an intrusive body of magma and Regional metamorphism is developed over extensive areas and is related to the formation of mountain ranges.

Some original rocks are subjected to pressure so intense or to heat so high that they are completely changed. The process of metamorphism does not melt the rocks, but can transform them into a denser, more compact foliated rocks (foliated means the parallel arrangement of certain mineral grains into layers). Unfoliated or nonfoliated metamorphic rocks, such as marble and quartzite, are also formed at this time. New minerals are sometimes created either by the rearrangement of mineral components or by reactions with fluids that enter the rocks. Contact metamorphism can occur today when molten lava is intruded into preexistent rocks. The rock adjacent to the intrusive lava is subjected to heat and pressure and undergoes metamorphism and can change into a different type of rock or another type of metamorphic rock. The metamorphic effect on three common sedimentary rocks would be as follows:

Limestone - Limestone is calcium carbonate ($CaCO_3$) which is easily affected by acids. When limestone deposits are subjected to heat and pressure they change into the metamorphic rock known as marble. The calcite minerals in the limestone are small, but grow larger and interlock when they become marble. Usually the heat and pressure destroys any remnants of the original limestone fossils, but sometimes the fossil evidence remains. Marble may have any color and is used in sculpture work and polished for floors, tabletops and interior and exterior walls.

Sandstone - When a quartz rich sandstone is subjected to metamorphism, it becomes a quartzite. The quartz grains have now become firmly bonded or fused by the entry of silica into the pore spaces. There is now limited pore space and the quartzite breaks right through the sand grains that make it up, rather than around them. Quartzite is typically very hard and resistant to erosion and weathering. One would normally find oil, gas or water in the pore space in sandstones, but little or none in quartzites.

Shale - Slate is a metamorphic rock that has been produced from low-grade metamorphism of shale or even clastic igneous rock. Additional heat or pressure would develop a series of rocks such as slate, phyllite, schist and gneiss (pronounced "nice"). Slate and phyllite are used for roofing and decorative floors. Slate can be any color, but black, green or maroon are the more common. The closer the rock or area is to a heat/pressure source, the higher its grade of metamorphism. Sometimes the aluminum in the clay or shale will react to form corundum or beautiful garnet crystals, especially when the metamorphic rock develops into a schist. The famous garnet crystals in a schist at Wrangell, Alaska are an example.

BIBLICAL APPLICATIONS

One might wonder about the type of rock/stone tablets on which the Ten Commandments were carved. Nearly in the center of the Sinai Peninsula, which stretches between the Horns of the Red Sea, lies a wedge of granite and other igneous rocks rising to between 8,000 feet and 9,000 feet above sea level. These features/deposits are possible igneous intrusive bodies. The description of Mount Sinai in Exodus 19:18 could be the power of God or volcanic activity.

> All Mt. Sinai was covered with smoke because Jehovah descended upon it in the form of fire; the smoke billowed into the sky as from a furnace, and the whole mountain shook with a violent earthquake (Exodus 19:18 LB).

All the above descriptions are volcanic type actions. While Moses was on Mount Sinai God gave him two tablets of stone on which the Ten Commandments were written with the finger of God (Exodus 31:18). They were breakable because Moses threw the tablets down in anger and they lay broken at the foot of the mountain (Exodus 32:19). They were local stone because God said later to prepare two stone tablets like the first ones and I will write upon them the same Commandments that were on the tablets you broke (Exodus 34:1). Moses took the two local stone tablets and climbed Mount Sinai. Several types of igneous rocks would be available, but granite would be very heavy and an added difficulty, while climbing mountains. It could be that a lighter material, like pumice or scoria, was used and not granite. The size of the tablets is unknown, however, and they might have been quite small. All types of igneous rocks would be present, however, due to the volcanic nature of Mount Sinai. Tablets of pumice or scoria would be breakable (yet strong), lightweight and easier to carry while climbing a mountain. A lightweight igneous rock is a good possibility, but that doesn't rule out the other types completely.

Rock is often mentioned in the Bible. A natural occurring inorganic substance in its natural habitat would be a rock. If removed and used for something it would become a stone. So the large massive fortress at the Strait of Gibraltar would be the Rock of Gibraltar, not the Stone of Gibraltar. Ayers Rock in Australia is correct because it's large and in place. The "Dome of the Rock" in Jerusalem is built over Mt. Moriah, which can be seen inside the mosque. Moses was once told by God to speak to a rock and it would pour out water for the people and cattle (Numbers 20:7-8). He struck the rock twice and water gushed out. Because of his disobedience (not obeying exactly), he was not allowed to go into the Promise Land.

Stones are rocks that are used for something. In the Bible, rocks were removed to stone people to death. Therefore, we have gemstones, cornerstones, building stones and the special Blarney Stone at the Blarney Castle in Ireland.

Rock layers are good foundations and considered a safe place to be. God is our foundation and our Rock of Ages. God is the rock that cannot be moved, so if we stand on Him and His promises, we will not be moved. Psalm 62 stresses God is the rock who can rescue.

> I stand silently before the Lord, waiting for Him to rescue me. For salvation comes from Him alone. Yes, He alone is my Rock, my rescuer, defense and fortress. Why then should I be tense with fear when trouble comes? (Psalm 62:1-2 LB).

> He only is my rock and my salvation: He is my defense; I shall not be moved (Psalm 62:6 KJV).

During Jesus Christ's ministry some said He was John the Baptist, some Elijah, some Jeremiah or one of the prophets. Peter was asked by Jesus in Matthew 16:15, "Who do you think I am?" Peter answered in Matthew 16:16 LB, "Thou art the Christ, the Messiah, the Son of the Living God." Peter's name had been changed by Jesus to Cephas, meaning a stone. Peter's testimony that Jesus was the Messiah, the Son of the Living God was the solid rock statement of all times. This was personally revealed to Peter by God the Father (Matthew 16:17). Upon this immovable rock statement of absolute truth the Son of God will build His church and the gates of hell shall not prevail against it (Matthew 16:18). We are to come to Christ who is the living foundation of rock upon which God builds (I Peter 2:4). We can then say like the Psalmist in Chapter 62...He (Christ) alone is my rock and salvation and, also, If God be for us, who can be against us (Romans 8:31). If we stand on the promises of God, believe Him with all our heart, mind and soul, confess Him openly and obey His commandments, we receive His spirit and are born again.

We then become children of God, who He loves. We are then under His total everlasting protection and nothing can separate us from His love.

> For I am persuaded, that neither death, nor life, nor angels, nor principalities, nor powers, nor things present, nor things to come, nor height, nor depth, nor any other creation, shall be able to separate us from the love of God, which is in Christ Jesus our Lord   (Romans 8:38-39 KJV).

Now, Jesus is the Rock of Ages and we (believers) become building-stones for God's use in building His house (I Peter 2:5). Jesus is precious to those who believe, but rejected and ridiculed by many. Even though He is rejected and overlooked by the majority, He has become the most honored and important part of the Church's building, the Cornerstone (I Peter 2:8). On this Cornerstone, we read what the building (church) stands for and to whom it is dedicated. The non-believers will see the truth and stumble over it. The beauty of creation and intelligent design and Biblical revelation are irrelevant to them. Their rejection will be their downfall. The non-believers will stumble because they will not listen to God's Word, nor obey it, so punishment must follow...that they will fall (I Peter 2:8). Jesus is a rock of offense to the non-believer. Believing in God as the Creator and the Cross being His only Salvation Plan, would brand natural man a hell-deserving sinner. His only hope would be the acknowledgment of Christ, the rejected Creator, as his personal Savior, who died for his sins.

Jesus is the living stone (not just a foundation rock) because He brings life and sustenance. He was chosen a Cornerstone long ago.

> But the Lord God says, See, I am placing a Foundation Stone in Zion - a firm, tested, precious Cornerstone that is safe to build on (Isaiah 28:16 LB).

We can now know that the church can never fall because of its foundation. Paul states it's the only foundation to build on (I Corinthians 3:11). No one can ever lay any other real foundation than that one we already have - Jesus Christ (I Corinthians 3:11). We can accept this rock or stone as our everlasting foundation or reject it and trip and stumble over it.

Jesus told a story in Matthew 21:33-41 representing the Jews' rejection of Him as the Son of God. He quoted them (Psalm 118:22), "The stone (Christ) which the builders refused is become the head stone of the corner."

Since they (the Jews) rejected Him, "The Kingdom of God shall be taken from them and given to a nation bringing forth fruits thereof and whosoever shall fall on this stone shall be broken; but on whomsoever it falls, it will grind him to powder" (Matthew 21:42-44; see also Romans 9:33, Acts 4:11, Zechariah 3:9).

The old hymns tell the story. In "The Church's One Foundation" we see these words, "The Church's One Foundation" is Jesus Christ her Lord. Also, "How Firm a Foundation" and "Rock of Ages...cleft for me, let me hide myself in Thee" and also the hymn which states, "On Christ the Solid Rock I Stand, All Other Ground is Sinking Sand."

Those who have rejected Christ...denying God's existence and power and promoting evolution will some day realize the end times are near. The kings, great men (world leaders), rich men and the Chief Captains' mighty men and every bondman (all men great and small) will hide themselves in the rocks of the mountains (Revelation 6:15). They will cry out for the mountains and rocks to fall on them to escape the wrath of God (Revelation 6:16). Only God can speak to a rock and have it obey, however. It will be too late as each one has already chosen his fate. The unbelieving and all liars (and evolution is a lie) shall have their part in the Lake of Fire (Revelation 21:8). The born again, obedient believers will be welcomed into Heaven and we will have wonderful new bodies in Heaven, homes that will be ours forever, made for us by God Himself and not by human hands (II Corinthians 5:1). Christians will also be given a white stone, probably a diamond, on which is engraved our new heavenly names (Revelation 2:17).

CHAPTER 4

MINERALS

Minerals are naturally occurring, crystalline, inorganic substances. They differ from rocks in that they have a definite chemical composition and certain physical properties. There are about 3,000 named minerals known on the Earth's crust. Most minerals are crystalline substances; that is, they have an orderly interval structure (arrangement of atoms). Minerals are a form of matter since they are material substances that occupy space. Matter exists in three states, which are gaseous, liquid and solid. All matter appears to be essentially electrical in nature. The crust of the Earth, its rocks and minerals are formed of elements. Eighty-eight elements have been discovered in nature and at least 15 have been made in the laboratory.

An atom is the smallest unit of an element. Each element is a special combination of protons (positive charged particles), electrons (negative charged particles) and neutrons (uncharged particles). Most of the mass or weight of an atom is in its nucleus or center. The nucleus is composed of protons and neutrons (except for hydrogen). The nucleus is surrounded by a varied amount of negatively charged electrons. The nucleus of a normal atom contains as many protons and neutrons as the surrounding cloud of electrons. Atoms are invisible to the eye and Hebrews 11:3 LB states, "All things...were made at God's command; and that they were all made from things that can't be seen." God made the atoms and then possibly used various combinations for all substances, elements and minerals. All matter was created by God, "In the Beginning..." (Genesis 1:1).

Minerals

Dr. Henry M. Morris in *The Genesis Record*, Page 41, writes:

> Not only does the first verse of the Bible speak of the creation of space and matter, but it also notes the beginning of time. The universe is actually a continuum of space, matter, and time, no one of which can have a meaningful existence without the other two. The term *matter* is understood to include *energy*, and must function in both space and time. - Thus, Genesis 1:1 can legitimately and incisively be paraphrased as follows: "The transcendent, omnipotent God-head called into existence the space-mass-time universe."

Some minerals are composed of only one element such as diamond (C), copper (Cu), silver (Ag), gold (Au), mercury (Hg) and sulfur (S). The abbreviations like (C), (Cu) and (Ag) are the chemical symbols for these elements. Although there are about 3,000 named minerals, the basic ingredients of rock and minerals are few. Virtually 99 percent of the Earth's crust is composed of eight unequally distributed elements and oxygen alone accounts for about half of the crust's weight (see TABLE 4.1).

TABLE 4.1 – MAJOR ELEMENTS OF THE EARTH'S CRUST

ELEMENT	WEIGHT (PERCENT)	ATOM* (PERCENT)	VOLUME* (PERCENT)
Oxygen	46.60	62.55	93.77
Silicon	27.72	21.22	0.86
Aluminum	8.13	6.47	0.47
Iron	5.00	1.92	0.43
Calcium	3.63	1.94	1.03
Sodium	2.83	2.64	1.32
Potassium	2.59	1.42	1.83
Magnesium	2.09	1.84	0.29
TOTALS	98.59	100.00	100.00

*Computed as 100 percent, hence approximate. After Brian Mason, *Principles of Geochemistry*, John Wiley & Sons, Inc., 3rd ed., 1966

Since silicon and oxygen are so abundant, we can see why more than 90 percent of the rock forming minerals are silicates (compounds containing silicon and oxygen and one or more metals). The most common of the rock forming silicate minerals are olivine, augite, hornblende, mica's, feldspar and quartz.

A beautiful feature of minerals is that if it grows or develops without interference, it will form a characteristic crystal form. This unique shape is an external expression of its internal arrangement of atoms. The crystal form of a mineral is critical in mineral identification as each mineral has a specific crystal shape. Minerals are beautiful creations of God, like wildflowers in nature. Mineral crystals are so magnificent in their beauty and symmetry; it is sometimes hard to believe they are natural. Besides crystal shape, other properties used to identify minerals are

Minerals

hardness, specific gravity, cleavage, color, streak and striations. Hardness is a very important property. Mohs' scale of hardness is based on assorted minerals representing various hardness from one to ten. This scale was devised by the German mineralogist Frederick Mohs (1773-1839), who selected these ten minerals because they are common or readily available (with the possible exception of diamond, hardness 10). Each mineral in the Mohs' scale can scratch any mineral with the same number or a lower number, and can itself be scratched by any mineral with a higher number. The selected minerals are as follows:

> No. 1 -- Talc, softest mineral
> No. 2 -- Gypsum
> No. 3 -- Calcite
> No. 4 -- Fluorite
> No. 5 -- Apatite
> No. 6 -- Orthoclase
> No. 7 -- Quartz
> No. 8 -- Topaz
> No. 9 -- Corundum
> No. 10 -- Diamond, hardest mineral

Minerals were possibly created for man's use in Genesis 1:1-2 when much volcanic activity might have been present. Uplift and mineralization would have also occurred on the third day of creation (Genesis 1:9-10), when the dry land was uplifted forming the primeval continents and primeval oceans. Dr. Henry Morris states in his annotations for *The DEFENDER'S Study Bible*, Page 5 (Genesis 1:9):

> The waters "under the heaven" apparently still contained all the material elements of the Earth in solution or suspension until the energizing Word initiated a vast complex

string of chemical and physical reactions to precipitate, combine and sort all the rock materials and metals comprising the solid Earth. The 'Earth' (Hebrew *eretz*) thus formed was the same 'Earth' which had initially been 'without form' (the same word *eretz* is used in Genesis 1:1,2,10), but it was now 'dry land,' no longer mixed in the initial watery matrix.

Beautiful mineral crystals can be found in rocks throughout the world. Very special minerals that are ornate and prized are called gems. Diamonds, emeralds, rubies, sapphires and topaz stand out as true gems. Other prized stones are classified as semiprecious or ornamental gemstones. Examples are quartz (citrine, rose, amethyst, agate), jade, turquoise, lapis-lazuli, garnet and tourmaline. Scarcity and fashion are very important in determining the value of a gem, but the following physical properties are prized: luster, transparency, color and hardness.

Some people restrict the use of "gemstone" to shaped and polished minerals and rocks used for personal adornment. This would omit, however, the organic substances including pearls, shells, amber and coral. Clearly, there are no precise limits to the term "gemstone;" and collections of gems may well include carvings, spheres and boxes fashioned from minerals, rocks and organic substances.

BIBLICAL APPLICATIONS

God knew that special minerals would be prized and used by mankind. Therefore, the Garden of Eden was planted close to the Pison River, a branch of the river that flowed out of Eden. Nuggets of gold, onyx and possibly lapis-lazuli were found in this area (Genesis 2:11-12). Man was created so intelligently that, in very ancient times, he was able to quarry some metallic minerals and produce brass and iron (Genesis 4:22, Job 28:2). Modern archaeology is now confirming the high degree of technology associated with the earliest human settlers all over the world.

Job 28:1, 6 KJV mentions the following about early mining:

> Surely there is a vein for the silver, and a place for gold where they find it. The stones of it are the place of sapphires: and it hath dust of gold.

The priceless worth of wisdom is illustrated by declaring it more valuable than gold and silver (Job 28:15) or precious onyx, the sapphire and jewelry (Job. 28:16-17). It is also more valuable than coral, pearls and rubies and the topaz of Ethiopia (Job 28:18-19). Where does one obtain this great gift of wisdom? "Behold, the fear of the Lord, that is wisdom; and to depart from evil is understanding" (Job 28:28).

The Jews were told the Promised Land was located where stones are iron, and out of whose hills thou mayest dig brass or copper (Deuteronomy 8:9). At Ezion-geber archaeologists have found ruins of Solomon's smelters, furnaces, crucibles, and refineries; also, copper and iron ore deposits (*Halley's Bible Handbook*, Page 183).

During the exodus from Egypt, Moses was instructed to take an offering to build the Tent Tabernacle. Gifts could include gold, silver, bronze, onyx stones and stones to be set in the Ephod (Exodus 25:1-7).

Later, the Ark of the Covenant was overlaid with pure gold (Exodus 25:11) and the mercy seat was made of pure gold (Exodus 25:17), as well as the dishes (Exodus 25:29). Gold, sardius, topaz, emerald, sapphire, diamond, agate, ligue (possibly amber), amethyst, beryl, onyx and jasper, all set in gold, were used to make the priestly robes (Exodus 28:9-20).

A description of the Holy City relates that jasper, sapphire, chalcedony, emerald, sardonyx, sardius, chrysolite, beryl, topaz, chrysoprasus, jacinth, amethyst and pearls adorn this beautiful City. The City contains gold so pure it's clear as glass. This clear gold is so abundant that it's used as street material (Revelation 21:18-21). Only those whose names are written in the Lamb's Book of Life are allowed to enter (Revelation 21:27). And whosoever was not found written in the Book of Life was cast into the lake of fire (Revelation 20:15).

Man has polluted this world so badly it will have to be destroyed (II Peter 3:10-12). New Heavens and a new Earth, wherein dwelleth righteousness (II Peter 3:13), will be created. God can then dwell with those who chose to honor Him. The Holy City will glow with God's glory (Revelation 21:23) and will be so pure it will shine like a precious gem (Revelation 21:10-11). "For the Glory of God did lighten it, and the Lamb is the Light thereof" (Revelation 21:23). The beautiful, impure gems and minerals, present in today's sinful world, only give us a hint of how beautiful Heaven must be!

CHAPTER 5

FOSSILS

The study of fossils is called Paleontology. The word *fossil* is derived from the Latin word *fossilis* meaning "dug up." For many early years, any unusual object dug out of the ground was considered a fossil. Now, we only use those objects that are evidence of past life. There is much evidence to indicate that man has been interested in fossils since the very earliest times and fossil shells, bones and teeth have been found associated with the remains of primitive and prehistoric man.

Fossils are the remains or evidence of ancient plants or animals that have been preserved in the rocks of the Earth's crust. The best simple definition for a fossil is *any evidence of past life*. This evidence may be preserved in various ways. When most animals die, the body is rapidly destroyed usually by scavengers, bacteria, chemical decay, erosion or other geological agencies. Once-living material seldom lasts very long after death. Therefore, only a very few specimens escape destruction and become fossils. Yet, today we have found billions and trillions of fossils. This is evidence that most of the strata were deposited rapidly, not gradually (Otherwise how could so many fossils have been preserved?). It's as if a great flood caused the death and destruction of these mostly marine organisms. The original oceans were most likely very large and God said to marine organisms in Genesis 1:22, "Be fruitful, and multiply, and fill the waters in the seas." This was God's first command to living creatures (animals, unlike plants, possess consciousness). The marine organisms obeyed and the result was an abundance of marine animals that were later preserved as fossils. Since death did not enter the world until after man's sin (Romans 5:12), most of the fossils were laid down during the Genesis Flood of Noah's time.

Generally, if the ancient plants or animals had hard parts (such as bone, horn and shell) they are most likely to be fossilized. Also, if the body is rapidly buried, its preservation is more likely. This burial may bring physical or chemical changes that can also cause destruction. If the formation that contains the fossils is uplifted and exposed, weathering and erosion may also destroy the fossils. Thus, fossilization is a relatively rare event and the record of fossil life is incomplete. The incompleteness of the fossil record makes the reconstruction of past life very difficult. It has been estimated that only one out of 44 species of fossilizable marine invertebrates of the past is known. It is also estimated that of the 10,000 or so species found on a tropical riverbank, only 10 to 15 are likely to be preserved. Therefore, we will most likely never see all the many special creatures that God created in the past. Most all fossils have been found in sedimentary rocks. No fossils were discovered in the rocks returned from the moon, which were all igneous in nature.

Fossils form in the following ways:

1. Preservation - An example would be a hard part of an organism (like shells and coral for marine animals). With land animals, there would be teeth, bone, tusks and such. Sometimes, the land animal fossils are found in massive graveyards, containing fossils of different animals.

2. Petrifaction - After the death of an organism, the soft tissue is ordinarily consumed by scavengers and bacteria. Sometimes, if the organism is rapidly buried, ground water containing dissolved silica, calcium carbonate or iron may circulate through the enclosing sediment and be deposited in cavities once occupied by veins, canals, nerves or tissue.

If replacement or recrystallization occurs, this is petrifaction. The addition of a chemical substance into pore space is termed permineralization. When the original material is replaced with a new substance, it is termed replacement. This type of fossil usually retains the shape of the original organism. Sometimes even the fine growth lines on shells and delicate structures in petrified wood are accurately preserved. Petrified dinosaur bone is common in some areas.

3. Cast and Molds - Molds of fossils form if the fossil is dissolved leaving an empty space. Casts form if these molds are filled with minerals or sediments. Molds of tree trunks have been observed in Hawaii in lava flows that covered ancient forests.

4. Carbonization - All living organisms are composed of carbon, hydrogen, oxygen and water. Carbonization occurs if a fossil, such as a leaf, is squeezed and the fluids are removed, leaving a thin carbon film of the original organism. Worms and other animals without hard parts can be preserved in this way showing tiny delicate structures. Coal beds, made from the remains of plants, show many impressions of ancient vegetation.

5. Tracks and Burrows - Tracks of lizards, amphibians or dinosaurs are found throughout the world. Dinosaur nests with fossil eggs are also found in many locations (especially China, Western United States and Patagonia South America). Most of the dinosaur skeletons and nests have been found on the surface, or at a very shallow depth, indicating a more recent time frame. Fossil burrows of clams and worms are also found in sedimentary rocks.

6. Coprolites and Gastroliths - Fossil dung or excrement of animals (coprolites) has been found in rocks.

Sometimes this can tell much about the eating habits of ancient animal life. Gastroliths are smooth rounded pebbles found near the rib cage of dinosaur fossils. These pebbles most likely aided the dinosaur's digestion, just as gravel in the gizzards of chickens crushes grain.

7.   Frozen, Mummified or Preserved Soft Parts - Frozen mammoths and other northern animals have been uncovered with every part of the animal intact. Frozen fossils of humans (Inca children) have been found in the Andes Mountains (Johan Reinhard, National Geographic, *"Frozen in Time,"* November 1999, pp. 36-55) and the famous Iceman of the Alps was discovered in 1991 (David Roberts, National Geographic, *"The Iceman,"* June 1993, pp. 36-67). Mummification can also occur through the trapping and sinking of an animal in a tar seep or pit; the tar would inhibit the decay of the body parts. An example of this is the famous La Brea Tar Pits in Los Angeles, California. The peat bogs in Denmark have also yielded specimens (including man) with soft parts well preserved. Sometimes resin or amber entombs ancient insects and preserves perfect fossils.

Fossils are not mentioned in the Bible since it was the middle of the Fifteenth Century before the true origin of fossils was generally accepted. Before that time, fossils were explained as "freaks" of nature or related to some superstitious beliefs.

The author was once working on an oil well that tested some of the oldest of rocks called the Cambrian Age. Water was obtained from this formation and it was very salty. From this test, the opinion of this writer was that the original oceans or seas were salt water. It is also believed that marine animals were created to live in this environment. The oceans could have formed from the interaction of the original lava or magma and some of the water from vents and springs below the ocean. It is only in recent years that springs have been discovered on the ocean bottom (Richard A. Lutz, National Geographic, *Dawn in The Deep*, February 2003, pp. 92-103). In Job 38:16 LB God asked Job, "Have you explored the springs from which the seas come, or walked in the sources of their depths?"

BIBLICAL APPLICATIONS

As mentioned above, fossils are not recorded in the Bible since their nature was unknown until the Fifteenth Century. The greatest opportunity for abundant fossilization would have been the Genesis Flood where in a short period:

> And all flesh died that moved upon the Earth, both of fowl, and of cattle, and of beast, and of every creeping thing that creepeth upon the Earth, and every man: All in whose nostrils was the breath of life, of all that was in the dry land, died (Genesis 7:21-22 KJV).

If this had been a local flood, at least some of the animals and birds could have escaped to higher ground. Mount Ararat, where Noah's Ark rested after five months (Genesis 8:4), is 17,000 feet high and a logical site for the ark to rest. A flood that reached in excess of a 17,000 foot elevation is not a local flood. Marine fossils have been found on the summit of Mount Ararat, verifying the ocean was at that pinnacle at one time. Pillow lavas are also present on the summit of Mount Ararat and they are thought to be the result of underwater lava flows. This would lead one to believe that Mount Ararat, which is a volcanic mountain, might have formed during the early months of the Genesis Flood. This volcanic action, especially underwater activity (possibly mentioned in Genesis 8:2), could cause great earthquakes and large tidal waves or tsunamis.

These waves would not affect a boat at sea, but could be as high as 200 feet when they reach shallow water or the shore. They would create tremendous erosion and a great amount of sedimentation and devastation. The current geological formations, containing fossils and coal beds, would be the result of this great catastrophic event.

Psalm 29:3-10 seems to describe this great flood where ancient antediluvian forests were uprooted by rushing waters. This would form great mats of vegetation that would be deposited as coal beds and great petrified forests. Geologists had formerly thought these deposits would take long periods of time. That is, until the eruption of Mount St. Helen's in Washington State on May 19, 1980, when abundant sediments formed in one day. Dr. John D. Morris in *The Geology Book*, Master Books, p. 43 writes:

> Trapped inside this material were many trees which are now petrifying, being injected with the hot silica-rich waters flowing through these volcanic deposits.

Dr. Steven A. Austin found the following after several years of research at Mount St. Helen's:

RAPID FORMED STRATIFICATION
Up to a 400 foot thickness of stratum has formed since 1980 at Mount St. Helen's.

RAPID EROSION
The little "Grand Canyon of the Toutle River" is a one-fortieth scale model of the real Grand Canyon.

UPRIGHT DEPOSITED LOGS
An estimate of 19,000 upright stumps existed on the floor of Spirit Lake in August 1985. If found buried in the stratigraphic sequence, these trees might be interpreted as

multiple forests, which grew on different levels over periods of thousands of years.

PEAT LAYER IN SPIRIT LAKE

A layer of peat several inches thick has accumulated in Spirit Lake.

CONCLUSION

Mount St. Helen's provides a rare opportunity to study transient geological processes which produced, within a few months, changes which Geologists might otherwise assume required many thousands of years.

As the Genesis Flood waters assuaged or subsided (Genesis 8:1), other events might have occurred. Dr. Henry M. Morris writes the following opinion in the footnotes *of The DEFENDER'S Study Bible*, p. 24:

> As a result of the water subsiding, the phenomena described in Psalm 104:6-9 began to take place. The Earth's crust collapsed deep into the previous subterranean reservoir chambers, forming the present ocean basins and causing further extrusions of magmas around their peripheries and through openings in their floors. The light sediments in the sea troughs were forced upward by isostatic readjustments to form mountain ranges and plateaus.

Dr. Henry M. Morris and Dr. Martin E. Clark write in *The Bible Has The Answer:*

> It is significant that fossils, especially of large animals such as the dinosaur, must be buried

quickly or they will not be preserved at all. Furthermore, the sediments entrapping them must harden into stone fairly quickly, inhibiting the action of air, bacteria, etc., or else they will soon be decomposed and disappear. The very nature of fossilization thus seems to *require* castrophism - According to the Bible; death did not even "enter the world" until after Adam's sin (Romans 5:12). And the fossil record, more than anything else, is a record of death - in fact, of sudden death - and on a worldwide scale!

At the end of the creation period (Genesis 1:31), God pronounced everything in the whole universe "very good." Thus the struggling, groaning creation (Romans 8:22) everywhere evident in the fossil record must be dated Biblically as occurring after man's sin and God's curse on man's dominion (Genesis 3:17). And this can only mean that most of the sedimentary rocks of the Earth's crust, with their fossils, were laid down during the awful year of the great Flood, when "every living substance was destroyed which was upon the face of the ground" (Genesis 7:23).

The geologic column, rightly interpreted, therefore, does not tell of a long, gradual evolution of life over the geologic ages, but rather its polar opposite - the rapid extinction of life as a result of God's judgment on the antediluvians when "the world that then was, being overflowed with water, perished" (II Peter 3:6).

Gloria Clanin writes this opinion from *In The Days of Noah:*

> All of the land was under water, enormous earthquakes and undersea eruptions were continuing. During this time, great amounts of sedimentation occurred. The layers of mud containing "dead things" became the fossil and coal beds we have today.

The fossil organic matter was also the source for the world's oil and gas deposits.

SUMMARY

The great abundance of marine fossils in sedimentary rocks must mean they were laid down rapidly under flood conditions. The mixing of the fresh and salt water, plus tremendous wave action, would have killed abundant life at the conclusion of the Genesis Flood. All the eroded sediments would have been redeposited in great stratified formations. Tremendous continental uplifts could have taken place and great basins opened to receive the available sediments (Psalm 104:8). This uplift allowed fossils of marine organisms to be deposited on top of some of the tallest mountains. The waters of the Genesis Flood drained off into these new basins, scouring out great canyons, eroding still-soft sediments and laying down great thicknesses of alluvium in the newly formed valleys, flood plains and deltas. One can see from this study that the great catastrophe of the Genesis Flood appears to explain all the Earth's fossil and geological features better than evolution and the uniformitarian model. The March 2005 edition of *Science,* vol. 307, reported flexible blood vessels being found inside the thighbone of a *Tyrannosaurus rex.* This tends to show dinosaurs were living more recent and not 70 million years ago (Stay Tuned).

Further Reading and References:

Austin, Steven A., Ph.D., *Mt. St. Helens and Catastrophism*, Institute for Creation Research, Impact Article No. 157, July 1986.

Clanin, Gloria, *In the Days of Noah*, Master Books, 1999, p. 58.

Foster, Robert J., *General Geology*, Fifth Edition, Merrill Publishing Company, 1988, p. 303.

Morris, Henry M., Ph.D., *The DEFENDER'S Study Bible*, World Publishing, Inc., 1995, p. 24.

Morris, Henry M., Ph.D. and Dr. Martin E. Clark, *The Bible Has The Answer*, Master Books, 2001, pp. 109-110.

Morris, John D., Ph.D., *The Geology Book*, Master Books, 2000, p. 43.

CHAPTER 6

EARTHQUAKES

Earthquakes, perhaps more than any other phenomena, demonstrate that the Earth continues to be a dynamic planet, changing each day by internal tectonic forces. Most earthquakes occur along plate boundaries. As the plates move, these boundaries, spreading centers, subduction zones and transform faults will be the sites of the most intense earthquake activity.

Earthquakes occur during sudden movements along faults. During long periods of slow deformation, elastic strain builds up between the rock bodies on opposite sides of a fault. Slip along the fault is prevented by friction until a threshold of strain is exceeded. Then the rocks snap past each other along the fault to release some of the stored energy.

The process of rebounding generates vibrations called seismic waves. These vibrations are somewhat analogous to those produced by a pebble dropped in a pool of water. The stone introduces energy into the system, and concentric waves spread out in all directions from the point of disturbance.

Surface waves produced by an earthquake have been reported with heights of more than 0.5 meter and wavelengths of 8 meters. The period between the passages of wave crests can be as much as 10 seconds. The vibrations can continue for as long as an hour before the wave dies out. Three types of seismic waves are generated by an earthquake shock:

1. **Primary waves** (*P waves*), in which particles move back and forth in the direction in which the wave travels. They are the first waves to arrive at a recording station.

2. **Secondary waves** (*S waves*), in which particles move back and forth at right angles to the direction in which the wave travels.

3. **Surface waves** (Long or *L waves*), which travel only in the outer layer of the Earth, are similar to waves in water. They are the last waves to arrive at a recording station.

Surface waves (also called Large or Long Waves) cause the most damage during an earthquake. Together with secondary effects from associated landslides, tsunamis and fires, they result in the loss of approximately 10,000 lives and $100 million in property damage each year.

The point within the Earth where the initial slippage generates earthquake energy is called the *focus*. The point on the Earth's surface directly above the focus is called the *epicenter*.

The *intensity*, or destructive power, of an earthquake is an evaluation of the severity of ground motion at a given location. It is measured in relation to the effects of the earthquake on human life. Generally, destruction is described in terms of the damage caused to buildings, dams, bridges and other structures, as reported by witnesses. An intensity scale commonly used in the United States and a description of some of its criteria are presented in the Scale of Earthquake Intensity Chart on the following page.

SCALE OF EARTHQUAKE INTENSITY CHART

I	Not felt except by very few people under special conditions.
II	Felt by a few people, especially those on upper floors of buildings. Suspended objects may swing.
III	Felt noticeably indoors. Standing automobiles may rock slightly.
IV	Felt by many people indoors, by a few outdoors. At night, some are awakened. Dishes, windows and doors rattle.
V	Felt by nearly everyone. Many are awakened. Some dishes and windows are broken. Unstable objects are overturned.
VI	Felt by everyone. Many people become frightened and run outdoors. Some heavy furniture is moved. Some plaster falls.
VII	Most people are in alarm and run outside. Damage is negligible in buildings of good construction.
VIII	Damage is slight in specially designed structures, considerable in ordinary buildings, great in poorly built structures. Heavy furniture is overturned.
IX	Damage is considerable in specially designed structures. Buildings shift from their foundations and partly collapse. Underground pipes are broken.
X	Some well-built wooden structures are destroyed. Most masonry structures are destroyed. The ground is badly cracked. Considerable landslides occur on steep slopes.
XI	Few, if any, masonry structures remain standing. Rails are bent. Broad fissures appear in the ground.
XII	Virtually total destruction. Waves are seen on the ground surface. Objects are thrown in the air.

The *magnitude* of an earthquake is a measure of the amount of energy released. It is a much more precise measure than intensity. Earthquake magnitudes are based on direct measurements of the size (amplitude) of seismic waves, made with recording instruments, rather than on subjective observations of destruction. The total energy released by an earthquake can be calculated from the amplitude of the waves and the distance from the epicenter. Seismologists express magnitudes of earthquakes using the Richter Scale, which arbitrarily assigns 0 to the lower limits of detection. Each step on the scale represents an increase in amplitude by a factor of 10. The vibrations of an earthquake with a magnitude of 2 are, therefore, 10 times greater in amplitude than those of an earthquake with a magnitude of 1, and the vibrations of an earthquake with a magnitude of 8 are 1 million times greater in amplitude than those of an earthquake with a magnitude of 2.

The largest earthquake ever recorded (Chile, May 22, 1960) had a magnitude of approximately 9.5 (M_w – moment magnitude) on the Richter Scale. Significantly larger earthquakes are not likely to occur, because rocks are not strong enough to accumulate more energy. The approximate number per year of various Richter Scale Magnitude Earthquakes is shown on the chart on the following page.

RICHTER SCALE OF EARTHQUAKE MAGNITUDE

MAGNITUDE	APPROXIMATE NUMBER PER YEAR
1	700,000
2	300,000
3	300,000
4	50,000
5	6,000
6	800
7	120
8	20
>8	1 every few years

The primary effect of earthquakes is the violent ground motion accompanying movement along a fracture or fault. This motion can sheer and collapse buildings, dams, tunnels and other rigid structures. Secondary effects include landslides, tsunamis and regional or local submergence of the land. The following are a few examples of well-documented earthquakes in historical times.

The earthquake that shook Peru on May 31, 1970, had a Richter Scale magnitude of 7.8. Claiming the lives of 50,000 people, it was the deadliest earthquake in Latin American history. Much of the damage to towns and villages was caused by the collapse of adobe buildings, which are easily destroyed by ground motion. Eighty percent of the adobe houses in an area of 65,000 kilometers^2 were destroyed. Vibrations caused most of the destruction to buildings; but the massive landslides, triggered by the quake in the adjacent Andes Mountains, were a second major cause of fatalities. A huge debris flow buried 90% of the resort town of Yungay, with a population of 20,000, in a matter of a few minutes. Similar earthquake-triggered debris flows killed 41,000 in Ecuador and Peru in 1797 and, in 1939, killed 40,000 in Chile.

Another example of an earthquake at convergent plate margins is the 1960 earthquake, which occurred in the Andes Mountains, in Chile, and caused extensive damage from ground motion, landslides and flooding. It also produced a spectacular tsunami, which devastated seaports with a series of waves 7 meters high. This tsunami crossed the Pacific Ocean at approximately 1,000 kilometers per hour and built up to 11 meters high at Hilo, Hawaii. Twenty-two hours after the quake, it reached Japan, causing damage to property worth $70 million.

The earthquake that devastated southern Alaska late in the afternoon of March 27, 1964, was one of the largest tectonic events of modern times. Its magnitude was between 8.3 and 9.2, and its duration ranged from 3 to 4 minutes at the epicenter (by comparison, the San Francisco earthquake of 1906 lasted about 1 minute). The Alaskan earthquake was important to Geologists because it had such marked effects at the Earth's surface and because it was well documented. Despite its magnitude and severe effects, this earthquake caused far less property damage and loss of life than other national disasters (114 lives were lost and property worth $311 million was damaged), because, fortunately, much of the affected area was uninhabited. The crustal deformation associated with the Alaskan earthquake was the most extensive ever documented. The level of the land was changed in a zone 1,000 kilometers long and 500 kilometers wide (an area of 500,000 kilometers^2 was elevated or depressed). Submarine and terrestrial landslides triggered by the earthquake caused spectacular damage to communities, and the shaking spontaneously liquefied deltaic materials along the coast, causing slumping of the waterfronts of Valdez and Seward.

One of the most famous earthquakes in history was the San Francisco, California earthquake of 1906. It lasted only a minute, but had a magnitude of 8.3. The fire that followed caused most of the destruction (an estimated $400 million in damage and a reported loss of 700 lives). From a geological point of view, the earthquake was important because of the visible effects it produced along the San Andreas Fault. Horizontal displacement occurred over a distance of about 400 kilometers and offset roads, fences and buildings by as much as 7 meters.

The Tangshan quake, which shook China on July 28, 1976, was probably the second most devastating earthquake in history. (The most destructive also occurred in China, in 1556.) The enormous shock registered 8.2 on the Richter Scale. Although Chinese authorities have withheld many details about this disaster, scientists from Mexico and other countries were recently allowed to investigate the affected area. They say that Tangshan looked like Hiroshima after the explosion of the atomic bomb. The devastation of the city of 1 million was complete. The Tangshan quake killed 750,000 people.

BIBLICAL APPLICATIONS

Joe E. Lunceford, Professor of Religion, Georgetown College, Georgetown, KY, writes the following concerning earthquakes in *Eerdmans Dictionary of the Bible*, page 362:

Earthquake:

An unusual shaking or trembling of the Earth or some part thereof. Both Amos (Amos 1:1) and Zechariah (Zech. 14:5) refer to a literal earthquake (Heb. ra as) during the reign of Uzziah, king of Judah. Since these two prophets refer to it as "the earthquake," without any other identification, it must have been very destructive and familiar to their readers. Beyond its literal meaning, "earthquake" became a symbol of divine activity, especially in judgment (Isa. 29:6). The Earth, and Mt. Sinai in particular, are said to have quaked before the awesome presence of Yahweh (Exod. 19:18; Psalm 18:7 [MT8] = 2 Sam. 22:8, Ps. 68:8). The prophet Elijah apparently expected Yahweh to be present in an earthquake, though instead Yahweh appeared in a still small voice (1 Kgs. 19:11-12).

In the NT there is likewise a literal and a symbolic use of the term earthquake (Gk. *Seismos*). An earthquake is mentioned at Jesus' crucifixion (Matt. 27:51, 54) and at His tomb (28:2). An earthquake opened the prison doors for Paul and Silas after they had been incarcerated at Philippi (Acts 16:26).

The figurative use of earthquake as a symbol of God's activity is preserved in the Synoptic Gospels (Mark 13:8 par.), among several other symbols depicting the judgment of God at the end of the world. The author of the book of Revelation saw an earthquake as one of several symbols of God's judgment following the breaking of the sixth seal (Rev. 6:12), as evidence of God's activity in response to the prayers of the saints (8:5), and as the result of God's taking the "two witnesses" up into Heaven (11:13). The same writer saw an earthquake as following the appearance of the Ark of the Covenant in the temple (Rev.11:19) and, finally, as accompanying the judgments of the seventh bowl (16:17-21).

The earthquake mentioned in Amos 1:1 must have been very severe, for it was remembered for over 200 years. This earthquake during the time of Uzziah is also mentioned in Zechariah 14:5. This great splitting of the Mount of Olives, beneath which are known to exist even now many fault lines, will probably result from the global earthquake of Revelation 16:18-20. This future fault zone earthquake will make a valley running east to west and half of the mountain will move toward the north and half toward the south. The faults in this area, including the great faults that formed the Dead Sea Basin, are trending north and south as so stated in Zechariah. The west half of the Dead Sea Fault is moving to the south and the east half is moving to the north. This is a left lateral fault and has moved some 105 kilometers in the past. This north/south fault direction in this area was unknown prior to the 20th Century.

Just as stated in the Scriptures, God made all the rocks and set forth the tectonic features that allow earthquakes to form (John 1:3). Jesus, at the Triumphant Entry into Jerusalem, was asked by the Pharisees to rebuke the disciples' rejoicing. He stated, "I tell you that, if these should hold their peace, the stones would immediately cry out" (Luke 19:39-40). This Triumphant Entry on a donkey's colt was prophesied in Zechariah 9:9. The crucifixion of Jesus was too much for nature "...and the Earth did quake, and the rocks rent" (Matthew 27:51).

God used an earthquake for punishment when Korah, Dathan and Abiram incited a rebellion against Moses. This was while they wandered in the desert due to their disobedience. Moses spoke against them and the ground suddenly split open and swallowed them up, along with their tents and families and everything they owned. The Earth closed back up and they perished (Numbers 16:31-33).This is proof of God's total control of all of nature.

A similar incident occurred in the 1964 Alaskan earthquake. When the quake hit, 12-year-old Perry Mead III and other children went outside. Just as they reached apparent safety, a crevasse opened under Perry and his two-year-old brother Merrell, and they plunged from the sight of their brother and sister who stood on firm ground nearby. The bodies of the two brothers were never recovered (*Planet Earth, Earthquake*, Time-Life Books, Alexandria, Virginia, pp. 22-23). In Isaiah 29:6 LB the prophet mentions that God would use thunder, earthquake, whirlwind and fire against their enemies when they are faithful and obedient and rejoice in the Lord. The sad part was that the people at that time were denying God's creation as evolutionists do today. The Bible had already stated, "The fool hath said in his heart, There is no God" (Psalm 14:1). In the Living Bible, Isaiah 29:16 related the disbelieving people to a potter and his clay, How stupid can they be! Isn't he, the Potter, greater than you, the jars he makes? Will you say to him, "He didn't make us"? Does a machine call its inventor dumb?

Earthquakes

God can and has used all of nature to show His glory and fight for His followers. As previously stated, the prophet Isaiah disclosed that God would use thunder, earthquake, whirlwind, storm and fire (Isaiah 29:6) against an enemy. In Acts 16:26 God used a great earthquake to release Paul and Silas from prison..."And suddenly there was a great earthquake, so that the foundations of the prison were shaken: and immediately all the doors were opened, and every one's hands were loosed." The area around ancient Philippi, toward Istanbul, Turkey, is on a major fault zone (the North Anatolian fault) where abundant earthquakes occur today.

Christ gave us hints about the end times in Matthew 24:6-8, Mark 13:7-8 and Luke 21:9-11. The listed world wars and rumors of wars, famines, pestilence and earthquakes are common knowledge to all. Total world earthquakes have been gradually increasing in number since 1990. The earthquakes in many lands and places will be the beginning of the sorrows, signaling the return of Christ (Mark 13:8). The U.S. Geological Survey Earthquake Facts and Statistics website verifies this increase since 1990, and displays the following information for total world earthquakes as follows:

YEAR	TOTAL WORLD EARTHQUAKES
1990	16,612
1995	21,007
2000	22,256
2002	27,459

Only Luke's account includes "fearful sights and great signs shall there be from Heaven" among the signs as Christ's return is nearer. In recent decades such things as UFOs, space probes, moon landings, satellites, space platforms and great photographs by the Hubble Space Telescope and such could be the fearful sights or great signs from Heaven. Current telescopic photographs showing the majesty and grandeur of space is a great sign to the author. These wonders should immediately relate us to Psalm 8:3-4, "When I consider Thy Heavens, the work of Thy fingers, the moon and the stars, which Thou hast ordained; What is man, that Thou art mindful of him? and the son of man, that Thou visitest him?"

The entire marvelous complex of space/time/matter/energy is continually "uttering speech" and "showing knowledge," teaching men and women of all times and places that there is a great Creator God who made it all. Everyone should be able to look at the universe and realize, "The heavens declare the glory of God; and the firmament sheweth His handywork" (Psalm 19:1). The boundless space, the endless time, the infinite energies and the innumerable complexities of the matter of the universe all unite in irrefutable testimony to the God of creation. This is the great lesson engraved on the textbook of the universe for all to read and learn. The whole creation, indeed, declares the glory of God. God is actually known to everyone instinctly (Romans 1:19).

> Since the earliest times men have seen the Earth and sky and all God made, and have known of His existence and great eternal power. So they will have no excuse [when they stand before God at Judgment Day]. Yes, they knew about Him all right, but they wouldn't admit it or worship Him or even thank Him for all His daily care. And after awhile

> they began to think up silly ideas of what God was like and what He wanted them to do. The result was that their foolish minds became dark and confused. Claiming themselves to be wise without God, they became utter fools instead. And then, instead of worshipping the glorious, every-living God, they took wood and stone and made idols for themselves, carving them to look like mere birds and animals and snakes and puny men (Romans 1:20-23 LB).

Since these things (God in creation) should have been seen and understood by mankind from the very time of the creation of the world, it is clear that the latter did not take place billions of years before men appeared on Earth, as evolutionists and progressive creationists have alleged. Men and women have been in the world ever since its very beginning, and all should have recognized the reality of God, even before God gave His written revelation.

The beautiful heavens we can observe today will pass away when the Lord comes, as well as the Earth being burned up (II Peter 3:10). Before this happens, mankind will doubt.

> Knowing this first, that there shall come in the last days scoffers, walking after their own lusts, And saying, Where is the promise of His coming? for since the fathers fell asleep, all things continue as they were from the beginning of the creation (II Peter 3:3-4).

This holds true today. Most of the world's population is indifferent to God's return: In fact, most of the world's people do not even believe in a personal Creator God, and His plan of salvation. They are too busy "walking after their own lusts." Many believe the things that continue today, are the things that have always been and, therefore, always will be. This is the so-called principle of *uniformitarianism* detailed in Chapter 3, Rocks, which states, "The present is the key to the past." That what we observe was accomplished by the same natural processes that continue to operate today. This remarkable belief is known as evolutionary uniformitarianism and it dominates the scientific and educational establishments of every nation in the world today. This, indeed, is a most remarkable fulfillment of Peter's prophecy, and could indicate we are in "the last days."

Earthquakes are also mentioned in Revelation 6:12, 8:5, 11:13, 11:19 and 16:18. The noise and extreme weather conditions sometimes related to major earthquakes is especially evident in Revelation 16:18. This Revelation 16:18-20 earthquake will be the greatest since men were upon the Earth. There have been many great earthquakes on the Earth since the eruption of the primeval water channels that initiated the Great Flood and the tectonic uplifts that terminated it (Genesis 7:11, 8:2 and Psalm 104:6-9). This earthquake, however, is far worse; in fact, so terrible that every island fled away and the mountains were not found. Oceans or seas were present until Revelation 21:1 and an earthquake generated tsunami or tidal wave might contribute to this devastation. This will not be a destruction by water since this was a promise in Genesis 9:11. This will, however, prepare the Earth for the Great Kingdom age. The structures in all the cities will collapse except for Jerusalem. God will divide it through the midst of the Mount of Olives when the Lord returns to stand on the Mount of Olives. Jesus ascended from the Mount of Olives (Acts 1:10-12) and will return in like manner. His feet shall stand upon the Mount of Olives (Zechariah 14:4). The Old Testament prophets also foresaw this mighty earthquake (Isaiah 24:19-20, Joel 3:16, Haggai 2:6-7). The smoothing of the topography of the world in this way (earthquake activity) is in preparation for the millennial Earth as well as fulfilling Isaiah 40:4-5 where "...every mountain and hill shall be made low: and the crooked shall be made straight, and the rough places plain." The terrible earthquake of Revelation 16:18 will smooth out the Earth so that the Earth's lands will all be pleasantly inhabitable as in the beginning and before the cataclysmically change at Noah's flood. We can infer that the original geography was shallow seas and gently rolling or flat land since all was "very good" and was made for man's enjoyment and utilization (Genesis 1:26-28, 31).

The Earth once shook terribly at the time of the Flood, then again when its Creator died on the cross. But one greater still is yet to come... "so mighty an earthquake, and so great"...that "every island [will flee] away, and the mountains [will not be] found" (Revelation 16:18, 20).

The 2004 Indonesian 9.0 megathrust earthquake could be a sign for the second coming of Christ. Luke 21:25 LB lists roaring seas and strange tides (tidal waves) as part of the events to come.

FURTHER READING AND REFERENCES:

Freedman, David Noel, Editor-in-Chief, *Eerdmans Dictionary of the Bible*, Wm. B. Eerdmans Publishing Company, Grand Rapids, Michigan, 2000, p. 362.

Hamblin, W. Kenneth, *The Earth's Dynamic Systems*, A Textbook in Physical Geology, Macmillan Publishing Company, 1989, pp. 381-397.

Morris, Dr. Henry M., *The Heavens Declare The Glory of God*, Daily Devotionals, World Publishing, Inc., 1997, pp. 96, 134.

Morris, Dr. Henry M., *The DEFENDER'S Study Bible*, King James Version, World Publishing, Inc., 1995, pp. 753, 1230, 1406, 1455 and 1475.

Morris, Dr. Henry M. and Dr. Martin E. Clark, *The Bible Has The Answer*, Master Books, 2001, pp. 333-336.

USGS Earthquake Hazards Program: *Earthquake Facts and Statistics*, September 2, 2003, http://neic.usgs.gov/neis/eqlists/eqstats.html.

Walker, Bryce, *Planet Earth, Earthquake*, Time-Life Books, Inc., Alexandria, Virginia, 1984, pp. 22-23.

CHAPTER 7

SATAN AND CREATION

Before man and the Earth were created God had already created a heavenly host of spiritual beings. The nature and purpose of these "rulers" or "powers" are still a mystery today. All the unknown powers, thrones and such were created by God in Christ for Him.

> For by Him were all things created, that are in Heaven, and that are in Earth, visible and invisible, whether they be thrones, or dominions, or principalities, or powers: all things were created by Him, and for Him (Colossians 1:16).

Regardless of the many unknowns and unexplainable secrets in the Bible, all things were created by God, since only God can create (Romans 11:36, Ephesians 1:10 and John 1:3). Our finite minds cannot comprehend this normally unseen, invisible to us, spiritual world. "Oh, what a wonderful God we have! How great are His wisdom and knowledge and riches! How impossible it is for us to understand His decisions and His methods!" (Romans 11:33 LB).

We know of the existence of multitudes or legions of angels from scriptures such as Matthew 26:53, Hebrews 12:22, Psalm 68:17, Daniel 7:10, Luke 2:13 and Revelation 7:11. When the Son of Man shall come in all His glory and judge the world, He will bring His holy angels with Him (Matthew 25:31). We have little appreciation of how intimately angels are involved in our lives; to minister, (Hebrews 1:14) delivering us from danger (Psalm 34:7,

Psalm 91:11-12) and also to deliver us from our enemies (Psalm 35:4-6). Angels, like we, were created by God, simply to obey Him and be messengers and ministers (Psalm 103:20-21). Angels are mentioned in the Old and New Testaments, but only a few people, living or dead, have actually seen them. They are normally invisible to the human eyes. Occasionally, they have been seen by men such as in the days of Elisha (II Kings 6:17), at the birth of Christ (Luke 2:13), at His ascension (Acts 1:10-11) and at the tomb (Luke 24:4-8), as examples.

There is evidently an angelic hierarchy among the heavenly hosts. There are cherubim and seraphim (Genesis 3:24 and Isaiah 6:2), as well as Michael the Archangel (Jude 9). The Captain of these hosts is none other than the Lord Jesus Christ. In *Eerdmans Dictionary of the Bible*, p. 63, we find the following information:

> Angels are a part of the creation of God, created either in the beginning or sometime before the foundation of the Earth (Ps. 148:2-5; Neh. 9:6; Col. 1:15-17). They are of a higher order than humans (Heb. 2:7) and are greater in power and might (2 Pet. 2:11; cf. 2 Kgs. 19:35). However, they are not to be worshipped by humans (Col. 2:18; Rev. 22:8-9). Angels are not omniscient as is God, for they do not know the time of the coming of Christ (Matt. 24:36; cf. 1 Pet. 1:12). Neither are they omnipresent, for they are said to go from place to place (Dan. 9:21-23). Angels are spirit beings (Heb. 1:14). They do not die, nor do they marry (Luke 20:36; Mark 12:25). While the number of angels is never definitely given, they are said to be innumerable (Dan. 7:10; Heb. 12:22; Rev. 5:11).

Angels may be wicked (2 Pet. 2:4; Jude 6; cf. Rev. 12:7) or good. Good angels seem to operate in conjunction with the work of the Holy Spirit in bringing God's message to mankind. Should an angel bring another message, he would bear the curse of God (Gal. 1:8-9). Satan can still come as an angel of light, and his ministers as ministers of righteousness, in deceiving and powerful ways (2 Cor. 11:14-15). The phrase "thrones or dominions or rulers or powers" may refer to angels (Col. 1:16).

Angels have a part to play in God's order in Heaven, and archangels have certain responsibilities over other angels of their order (1 Thess. 4:16; cf. Rev. 12:7). Righteous angels perform the work of God (Ps. 103:20). They effect the vengeance and wrath of God upon the disobedient, as is seen in the scourge of Israel (2 Sam. 24:16); upon Balaam (Num. 22:31); and when Jesus shall come again with his mighty angels to take vengeance on those that do not know God or obey the gospel (2 Thess. 1:7-10).

With the above angelic information in mind, a theory follows that combines the spirit world, creation and current creationist geological knowledge. This narration is not provable and is only an opinion of what might have occurred in the past Earth and universal history.

When God began to create the world, all the angels were likely very interested in this exciting new event. Only God had the ability and knowledge to conduct such an important event. All things (in Heaven and in Earth), including the angels, had been created by Him, and for Him (Colossians 1:16). God had created all the angels, including Satan, as good spiritual free-willed beings. Satan was created as the highest of all of God's created angels (Anointed Cherub, Ezekiel 28:14) and placed at the head of the hierarchy. God might have told Satan of His plans for the upcoming universe. Angels, as stated earlier, were created as free spirits, not as unthinking machines. They, therefore, could reject God's will for their lives, if they so choose.

    Satan must have noticed, and most likely coveted, that God was speaking this fantastic creation into existence. The spectacular and grandeur nature of this marvelous event of creation must have been the talk of Heaven. All of Heaven were watching and observing a majestic new world being created and appearing in their presence. The creation formula $N = S$ (nothing equals something) could only be used by God Himself and He was also enjoying this good creation. It could be that God later used a reversal of this formula from $N = S$ to $S = N$ (something equals nothing) to forgive sins and remember them no more. He could then take the something or substance of sin (S) and when one properly confesses and repents (I John 1:9-10), God would then change the substance of this sin to nothing. God can remove sin so far away; it will be remembered no more (Jeremiah 31:34). He will remove forgiven sins as far as the east is from the west (Psalm 103:12). One could travel east (or west) forever without coming to an end. This is the limitless scope of God's forgiving grace.

    Satan and the other angels watched as God continued His creation. He created light on the first day and divided it from darkness. It must have been one of the sights of all sights. God's glory must have lit up the creation domain as it will someday in Heaven (Rev. 21:23). All were mystified at the creation of light, and its separation from darkness; allowing an evening and morning on the Earth's first day. The Earth was already spinning on its axis. The division of the waters above and below the sky was also a spectacular event for all to see on the second day. The angels and other powers possibly had many questions as to the purpose of this splendor; but could hardly wait for the next surprise event. On the third day, land and the first living organisms (plants) were created. All of Heaven enjoyed observing the manifest variety of plants that appeared on the Earth. Each plant was created with "functional maturity" and omnipotent intelligent design, so the plants were immediately blooming or bearing fruit. The variety, appearance and color of plant life were spectacular. Satan might have asked what the purpose was for vegetation and what would happen to the plants in the future. God might have stated that they would be food and nourishment for future creations, stop erosion, release oxygen and eventually wilt or die and enrich and become a part of the soil He had created them to grow in. The plants all looked beautiful and good enough to eat, but spiritual beings did not require food. The grandeur of the fourth day of creation (sun, moon, stars and the universe) was beyond description to all and they were continually amazed by the splendor and beauty of this heavenly event. God's creative glory and superb and palatial diversity were very impressive to every heavenly being. Water animals of an unbelievable variety followed and then an assortment of beautiful birds, on the fifth day. Sundry and grandiose land animals were created on the sixth day. Most of the living organisms were told to be fruitful and multiply and fill the waters or multiply on the Earth. God Himself enjoyed this creation very much and was pleased; "God saw that it was good" (Genesis 1:10, 12, 18, 21, 25).

Satan, the angel leader, might have become jealous of God's abilities and the attention and praise He was receiving. He (Satan) might have again inquired about the purpose and future of water animals plus cows, horses, dinosaurs and such of the land animals. God might have told him that these animals will eat the grass or plants He created earlier (that you asked about before) and are for the use or benefit of future creations. They will not be spiritual and eventually might die and become a part of the soil.

Well, the last or sixth day of creation found all of the heavenly hosts out to see the most unusual of all the creations. God and Christ were creating a man in Their image and after Their likeness, "Let Us make man in Our image, after Our likeness..." (Genesis 1:26-27). They (mankind) were to be the masters and have dominion of all life upon the new Earth, and in the skies and all the seas (Genesis 1:26). Man was made in the image of God, not of angels. Man was not only to possess a body (like plants and animals) and a consciousness (like animals) but a third entity: the image of God, an eternal spirit. He would, therefore, be capable of communications and fellowship with his Creator.

In *The DEFENDER'S Study Bible*, p. 7, Dr. Henry M. Morris writes an opinion about Genesis 1:26:

> Man was not only created in God's spiritual image; he was also made in God's physical image. His body was specifically planned to be most suited for the divine fellowship (erect posture, upward-gazing countenance, facial expressions varying with emotional feelings, brain and tongue designed for articulate symbolic speech-none of which are shared by the animals). Furthermore, his body was designed to be like the body which God had planned from eternity that He Himself would one day assume (I Peter 1:20).
>
> The "dominion" man was to exercise was to be over both "the Earth" and also all the other living creatures on the Earth. Such dominion

obviously was under God as a stewardship, not as autonomous sovereign. Man was to care for the Earth and its creatures, developing and utilizing the Earth's resources, not to despoil and deplete them for selfish pleasure.

And God blessed them (mankind), and God said unto them, "Be fruitful, and multiply, and replenish the Earth, and subdue it: and have dominion over the fish of the sea, and over the fowl of the air, and over every living thing that moveth upon the Earth" (Genesis 1:28).

No instruction was given to exercise dominion over other men, but only over the Earth and the animals. Had man not rebelled and sinned, the total world would have remained in perfect harmony and fellowship with God and, therefore, with one another. God gave man all the abundant fruits and plants to eat. He also gave plants for food to all animals (Genesis 1:29-30). Both mankind and animals were originally intended to be vegetarian or herbivorous. There was adequate nourishment and energy value in fruits and herbs to enable both to accomplish God's work for them. With lush vegetation everywhere, the animals soon populated all the Earth as God so planned.

When God completed the great panoramic majestic creation of the Earth and universe, all the heavenly hosts must have stood, clapped and shouted praises. God was also pleased and satisfied with everything He had made "...and behold it was very good" (Genesis 1:31). Then God rested on the seventh day. This was to be an example of working six days and then a day of rest. Today, this is the reason for the seven-day week. As God was resting, Satan again might have had some questions like: What about this man you have created out of dust or clay? (Genesis 2:7).

What will be his future? Will he just be food for others, and eventually die and become dust or dirt (clay) again?

God might have said to Satan, "I have chosen to make him immortal, with a spiritual body and to, therefore, live for eternity in Heaven. I look forward to fellowship with mankind."

Satan was very displeased and jealous and told the other angels about this in anger. When some heard that man, made from dust, would eventually live in Heaven, one might have remarked, "Well, there goes the neighborhood." Satan must have strongly disagreed with God's plan to make man immortal. He verbally protested and talked angrily to other angels about this matter and caused major problems in Heaven. Satan developed a following and he was also worried about losing his position in Heaven.

In *The Bible Has The Answer*, p. 307, we read:

> When God created him (Satan) and set him at the head of the angelic hierarchy, He undoubtedly told Satan of His plans for the universe, including the forthcoming creation of man in His own image. In view of Satan's later malevolent preoccupation with man's destruction, it seems that this intention of God provoked a spirit of resentment in Satan's mind. He developed an intense personal pride in this own exalted position, his beauty and wisdom, and was displeased that God would create a race of beings in closer fellowship with God than the angels and, furthermore, that he would give them the marvelous ability of reproduction and multiplication, a privilege not shared by the angels.

This spirit of pride and resentment then began to generate a spirit of unbelief in God's truth and sovereign righteousness. "After all," he probably reasoned, "how do I really know that God can do all He says? How do I even know that He created me? All I have is His word for it, and He probably just told me this to keep me in my place. He is probably no better than I am, except that He arrived on the scene before I did. It seems more reasonable that He also, like all the other angels, just arose by natural processes out of the elemental energies of the cosmos. With the advantage of His prior emergence and experience, He has been able to organize and control all the rest of us, and now, with this trick, He has learned of making men who can reproduce themselves, no telling where we may end up. The time has come to organize the other angels and institute a new order in the universe. I will be God myself!"

The resulting fall of man enabled sin to enter the world and Satan to usurp the dominion thereof. God has, in His inscrutable wisdom, allowed Satan to continue for a time in his rebellion, now with the Earth as his base of operations and with most of mankind as his unwitting allies. As time goes on, the conflict will become more open and intense. Even now there are millions who consciously worship Satan and many more millions who are increasingly open in their hatred of God. Ultimately the "world" will "worship the dragon" (Revelation 13:3, 4), but his success will be short-lived. For "the devil that deceived them was cast into the lake of fire...and shall be tormented day and night for ever and ever" (Revelation 20:10).

It is hard to imagine Satan's perseverance and deception methods, but one-third of the angels were misled and followed his rebellion in Heaven (Revelation 12:4-9, Ezekiel 28:14-17). Satan exalted his throne above the throne of God and stated, "I will be like the Most High" (Isaiah 14:13-14). This caused a great war in Heaven (Revelation 12:7) and Satan and the angels that followed him (later called demons) were cast out of Heaven into the Earth (Revelation 12:8-9). It could have been that prior to this "casting out" that Satan and his followers were captured by God for punishment. God might have told them they were upsetting His will, His domain and causing great harm in Heaven. It would be necessary, therefore, they be punished or eliminated. God might have asked them if they had any final words before He reversed His creation formula on them and made their being or something into nothing (S = N). Satan might have asked for a final word and he requested to be heard by all of Heaven. He might have said, "We know God, You probably have the power to eliminate us into nothing, but I wish to say this. You can see I'm not the only angel that feels the way I do. There are many thousands just like me" (Revelation 12:4, 9). "Also this special creation of yours - <u>man</u>: He has free will, he might also agree with me. If you were to allow me to ask, test or question him, he might join us also. You see, he has heard only one side of the story. I don't think its fair not giving mankind a choice."

In his great evil wisdom and created rhetorical ability, Satan actually challenged God before the heavenly tribunal. The other angels were shocked at such disrespect and brashness. Satan would use this same tactic later. In the Book of Job, we again see this ageless conflict between

God and Satan. God proudly talked about Job and his faithfulness and goodness to Satan. In this story we read the following:

> Then the Lord asked Satan, "Have you noticed my servant Job? He is the finest man in all the Earth - a good man who fears God and will have nothing to do with evil."
>
> "Why shouldn't he, when you pay him so well?" Satan scoffed. "You have always protected him and his home and his property from all harm. You have prospered everything he does - look how rich he is! No wonder he 'worships' you! But just take away his wealth, and you'll see him curse you to your face!"
>
> And the Lord replied to Satan, "You may do anything you like with his wealth, but don't harm him physically" (Job. 1:8-13 LB).

The Lord accepted this challenge from Satan in the book of Job. Dr. Henry M. Morris writes this opinion concerning Satan and his testing of Job in *The Heavens Declare The Glory of God*, p. 195:

> Still smarting with wounded pride that God would make His angels mere "ministering spirits" (Hebrews 1:14) to Adam and his children, whose own bodies were mere "houses of clay," built out of the dust of the Earth, these demonic rebels hate human beings - especially those who love and serve God - with great passion. If Satan could not destroy Job by tempting him into moral wickedness or rebellion against an "unjust"

God, perhaps he could lead him into discouragement, using his self-righteous "friends" to cause him to lose faith in God's love and care.

But he failed! Job said: "Though He slay me, yet will I trust in Him," and "I know that my Redeemer liveth" (Job 13:15; 19:25).

Such defeatism is one of Satan's most effective weapons. When he strikes with it, we must, like Job, "resist steadfast in the faith" (I Peter 5:9), knowing the end result will be good, "for He is full of tenderness and mercy" (James 5:11 LB).

Back to the creation story…With all of the heavenly beings looking on concerning the mankind challenge from Satan, God might have said, "Very well, we will see who man will follow. Mankind has free will and will be allowed to choose between us with a level playing field. I will allow you (Satan) and your followers (demons) freedom to live and keep your powers for a limited time or season. There will be some boundaries; I will not allow you to cross." (This was like saving or sparing Job's life, as occurred in Job 2:6.) "You Satan, must report in ever so often" (Job 1:6-7). At the end of your period of freedom, I will require justice and judgment. Those who follow Me will live in eternal glory and those who follow you will be sent to eternal punishment. This punishment is an eternal fire prepared for Satan and his demons (Matthew 25:41). Satan didn't believe God would eventually prevail. He believed that by lying, deception and a cunning nature, he would be the ruler of the Earth and battle with God when he had accumulated the greater forces.

Satan has lots of power today and is still a liar (as he lied to Eve in the Garden). Satan is the father of liars and a murderer from the beginning (John 8:44). The son of man came to destroy the works of the devil (I John 3:8). Satan is like a roaring lion, however, seeking whom he may devour (I Peter 5:8). He seems to actually be winning the majority of mankind's loyalty as we look at the wickedness (all centered in Satan) in the world. He is called the god of this evil world (II Corinthians 4:4) and prince of the air (Ephesians 2:2). Since Satan is the prince of the air, we cannot go out into the world without being contaminated, touched or tempted by this evil, satanic world. Our only defense in this spiritual warfare is being full of God's Spirit (Galatians 5:16) and putting on the whole armor (breastplate of righteousness, feet shod with the gospel of peace, shield of faith, helmet of salvation and the sword of the spirit of God). We must also watch and pray (Galatians 6:10-18).

Satan, demons and evil spirits are the cause of suffering, disease, frenzy, pestilence and even national calamities. As chief of a host of wicked spirits (Matthew 25:41), unseen forces and rulers of darkness (Ephesians 6:12), he stands behind and directs all nations and forces who oppress God's people. This would include terrorism. Satan's ultimate aim is to disrupt or destroy man's faithfulness to and his trust in God. He tests us by deception, stressing that what God has commanded us to do (to put our trust in Him) cannot be of God. Satan uses afflictions, persecutions and trials to attempt to turn the pious from whatever task God has laid upon them (I Peter 5:8-9).

> This same Satan seems also to have always been known as one who acts only within and under God's permissive will (note the coordination of divine and satanic purposes

> in Matt. 4:1) and even in his wickedness functions still as a divine servant, wittingly fostering in the world various aspects of God's righteousness. In Luke 22:31-34, e.g., Satan is described by Jesus as sifting out for God the impurities in the disciples' commitment. In I Cor. 5:1-5; I Tim. 1:20 Satan is one who can aid the sinner and the blasphemer respectively in returning to the fold (*Eerdman's Dictionary of the Bible*, p. 1170).

Satan is always looking for opportunities to mislead humans and even tried to obtain Moses' body for an added deception of God's people. He contended with Michael the Archangel over Moses' body and lost this battle when Michael said, "The Lord rebuke you" (Jude 9).

Satan had success with the first Adam in the Garden of Eden (Genesis 3:1-7). He was brash enough to probably ask for another testing opportunity with the second Adam or Christ; but this time he failed (Matthew 4:1-11).

Some of Satan's demons must have <u>committed</u> some extreme sins since they have been "reserved in everlasting chains under darkness unto the judgment of the great day" (Jude 6 and also II Peter 2:4). This could have been the attempt to corrupt all mankind by taking physical possession of "the daughters of men" to produce "giants in the Earth in those days" (Genesis 6:1-4, Job 4:18).

As previously stated, Satan was a murderer from the beginning and is the Father of Lies (John 8:44-45). He told the first lie on Earth to Eve in the Garden of Eden (Genesis 3:4-5). Dr. Henry M. Morris writes in *The DEFENDER'S Study Bible* (p. 1149), the following, concerning Satan, as recorded in John 8:44:

> 8:44 *murderer from the beginning.* Evidently, the first child of Satan was Cain who slew his

brother Abel, no doubt at the instigation of Satan (I John 3:8-12). Since that first murder, the devil has been seeking to slay men and women before they can become children of God through faith in Christ, using his own children whenever he can to accomplish it.

8:44 *father of it*. As the father of lies, the devil deceived our first parents with the lie that they would become "as gods" (Genesis 3:5) through obeying his word rather than God's Word (Genesis 3:1-5). This lie of "humanism" - that men and women, as the apex of the evolutionary process, are the true gods of the world - has been deceiving and drawing people away from the true God of creation ever since. It has assumed various forms in different times and places, but it is always essentially the same old lie of Satan "which deceiveth the whole world" (Revelation 12:9). Thus, he is author of the great lie of evolution, seeking to understand and control the world without its Creator. He has, thereby, deceived himself first of all, convincing himself that both he and God had evolved out of the primeval chaos (as in all the ancient mythical cosmogonies which he must have taught his own earliest human children).

Evolution (leaving God out of creation and history) is the great lie of this century. Most of the colleges teach a godless creation of the Earth. The instructors who teach this have much head knowledge and little wisdom, "The fear of the Lord is the beginning of knowledge: but fools despise wisdom and instruction" (Proverbs 1:7 and also Proverbs 9:10). They need the great wisdom of Solomon who, after much study, declared in Ecclesiastes 12:13..."Let us hear the conclusion of the whole matter: Fear God, and keep His commandments: for this is the whole duty of man."

All who believe and trust in a godless evolution believe and practice a lie; and all liars, murderers, unbelieving and such, are thrown into the Lake of Fire (Revelation 21:8). This includes Satan (Revelation 20:10).

Evolution and Christianity are both based on faith. Only one idea is correct. By faith, we understand that God created the Heaven and the Earth (Hebrews 11:3). We will never know the *how* of the Earth's creation, but we know the *why*. It is also impossible to please God without believing God exists and that He is the Creator of all things (Hebrews 11:6)

We must realize that a total and complete acceptance or surrender and belief in the Creator are necessary for salvation. This story is paraphrased in John 3:16:

> For God (the greatest Creator) so loved (the greatest of all love) the world (the greatest creation) that (the greatest conjunction) He gave (the greatest gift) His only (the greatest sacrifice) begotten Son (the greatest being) that whosoever (the greatest group) believeth (the greatest necessity) in Him (the greatest Deliverer), should not (the greatest reason) perish (the greatest disaster), but have (the greatest hope) everlasting life (the greatest gift).

One must believe with all his heart, mind and soul to receive God's favor and salvation and to be born again. "Verily, verily, I say unto thee, Except a man be born again, he cannot see the Kingdom of God" (John 3:3).

Satan led the now-fallen angels into a cosmic rebellion against their Creator, which continues to this day. This must have been the origin of evil, an opposing force to God. The fact of evil powers is known only too well to each of us - we have only to look back to our own hearts before we were born again, to acknowledge that. We can also feel his daily temptations.

God is the origin of goodness and Satan, the origin of evil. Satan tempts mankind to sin, be enticed and lust after the flesh. God never tempts us to commit evil (James 1:13-14). Satan had a temporary victory when mankind disobeyed and sinned in the Garden of Eden (Genesis 3:1-7). God punished Adam and Eve for sinning and Satan assumed he had won the allegiance of the woman and all her descendants. God told Satan, however, there would be enmity between him and the woman,

> ...and I will put enmity between thee and the woman, and between thy seed and her seed; it shall bruise thy head, and thou shalt bruise his heel (Genesis 3:15).

Later, an animal was slain to make clothing for Adam and Eve (Genesis 3:21). Dr. Clifford A. Wilson writes in *Questions and Answers on Creation*, p. 30, concerning this:

> Before the clothing could be given, the animal had to die. Before the sinner could become the saint clothed in Christ's robe of righteousness, the Lamb of God had to die. And that was when the Seed of the woman had His heel bruised. That was when the serpent's head was crushed. Satan hoped that Calvary would be his ultimate triumph and the final defeat of the Son of God. Instead, at

this great battleground, the victory of the Son of God was complete and paradise could be restored.

This prophecy was fulfilled in the first instance at the cross, but will culminate when the triumphant Christ casts Satan into the Lake of Fire (Revelation 20:10).

Today, Satan and his followers lead the charge of evil with the popular secular humanism, new age movement and evolutionary teachings prevalent in today's society. They have become associated with both the sky-images of astrology and the corresponding graven images of paganism. Materialism is the great idol god of our times. Paul warned that such idol worship was, in reality, demon worship (I Corinthians 10:20). It is this particular "host of Heaven" which all devotees of false religions, ancient and modern, have really worshipped, when they reject the true God of creation and put their faith in some aspect of the cosmos itself. The faithful and obedient host of Heaven worships God alone, and so should we.

FURTHER READING AND REFERENCES:

Freedman, David Noel, *Eerdmans Dictionary of the Bible*, William B. Eerdmans Publishing Co., Grand Rapids, Michigan, 2000, pp. 63-64, 1149, 1170.

Morris, Dr. Henry M., *The DEFENDER'S Study Bible*, World Publishing, Grand Rapids, Michigan, 1995, pp. 3-9, 13.

Morris, Dr. Henry M. and Dr. Martin E. Clark, *The Bible Has The Answer*, Master Books, 1987, pp. 306-307.

Morris, Dr. Henry M., *The Heavens Declare the Glory of God*, Daily Devotionals, World Publishing, Inc., 1997, pp. 195, 242.

Wilson, Dr. Clifford A., *Questions and Answers on Creation*, Pacific Coast Ministries, 1994, pp. 23-30.

CHAPTER 8

CONCLUSION

    One looks at the present world and sees a great variety of religions and ideas concerning what to believe. Everyone should seek the capital "T" truths about what we should believe, and put our confidence in. Little "t" truths, such as water is wet, the sky is blue and grass is green, are readily accepted by everyone. But the most important question one can ask and solve is this: Is there a God, who or what is it and what does this God (god) require? This is the most important question in life and the solution is a most valuable treasure. Who can we trust to teach us the correct answer to this most important question? Most people adopt the religion of their families. I feel that everyone should examine his or her inherited family religion to see if it is really truth. There is much variety and numerous types of religions; however, they cannot all be correct. What or who are the all important true God and Creator of this world? How can a person be sure he is on the right pathway to acquire the wisdom and knowledge necessary to find the one and only true God?
    The author believes the Bible is truth and contains all the knowledge and wisdom needed for enlightenment and revelation. If one doesn't agree with this assumption, then someone is wrong, because both opinions cannot be correct. A person would only find the correct final answer when he died and faced eternity. This eternity would either be nothingness or facing a Holy and just God. It is then too late to change, if you chose unwisely. The Bible states that all idolaters and unbelieving shall have their part in the Lake of Fire (Revelation 21:8). It is, therefore, very important to

find truth before death. To be certain, one must seek to find the correct answers. Matthew 7:7-8 admonishes us to ask, seek and knock; and if we do so, we will find. One should never be afraid to examine closely the religion or philosophy we now possess. Most people take the word of a parent, minister, priest, professor, author, friend or neighbor concerning their eternal destiny. What if the person you trusted is not correct? If someone saw a person dig out a rock or cut down a tree and shape it into a god, he should be skeptical of that god. We know this god was created by the worshiper and really has no power. Isaiah 40:18-20 describes this practice of a self-created god who cannot even move, much less talk or give empowerment. God warned about this practice in Exodus 20:4-5.

Could it be that some other god or religious practice is of a higher order and more powerful than the Jehovah God of the Bible? Could the Hindus, Buddhists or others have the real true god? One should always check and see if any religion is worshiping creation instead of a Creator. This would be like man-made idols, sun gods, astrology, materialism or ritualistic type gods or religions. Can any of creation (sun, moon, stars, trees, and rocks) talk or have god like power? Are any of these gods a true and living God, like the God of the Bible?

One should consider, therefore, the Holy Bible, which is the inspired Word of the Living God. This book is free of error, scientifically correct and has survived thousands of years of testing and trials. The Bible is the "Word of God and Jesus Christ is the Living Word" (John 1:1-3, Revelation 19:13). Both Words are without error; both are divine, yet can be comprehended by man. Science books change data each year, but the Bible never changes, since it is the Word of God. "The grass withereth, the flowers fadeth; but the word of our God shall stand for ever" (Isaiah 40:8 KJV). The principles, ideas and plans of Jesus are always the same. "Jesus Christ the same yesterday, and today, and forever" (Hebrews 13:8 KJV).

One can really see God in creation displayed by intelligent design. God created all things. Do you know any other gods or religions that state that they created the heavens and earth, like the God of the Bible (Genesis 1:1, Exodus 20:11, John 1:3, Colossians 1:16)? God can only speak truth and He stated that He created all things. Only the Bible tells the *why* of creation.

A person can choose not to believe in God and His Holy Word, the Bible. This professing to not believe in God changes nothing. He either exists or the Bible and His Word are a lie. It would be very foolish to determine, with all the visible evidence, that there is no God. "The fool hath said in his heart, There is no God" (Psalm 14:1). These so called atheists usually have abundant worldly knowledge and yet lack Godly wisdom. This worldly wisdom or knowledge will someday be destroyed (I Corinthians 1:19). Nonbelievers are usually very prideful. "The wicked, through the pride of his countenance, will not seek after God: God is not in all his thoughts" (Psalm 10:4 KJV). The knowledge of this world will never comprehend the creation and construction of this complicated world. No solution to the exact nature of creation, living beings, the human body, universe, plants and animals has occurred from extensive modern scientific research. God knows all the answers and His ways and thoughts are beyond our ways (Isaiah 55:9). He, therefore, is the only God (Isaiah 43:11). If God had a weakness, we are too finite to find it (I Corinthians 1:25). It is impossible to determine the nature of an infinite God with a finite mind. If we had all the knowledge in the world we still might not be safe in eternity. What profit is there if you gained the whole world and lost your soul? (Matthew 16:26). That would be the greatest of all tragedies. One must have wisdom and understanding to know an invisible omnipotent God.

Conclusion

"How does a man become wise? The first step is to trust and reverence the Lord!" (Proverbs 1:7, 8, 9 LB).

> For the reverence and fear of God are basic to all wisdom. Knowing God results in every other kind of understanding (Proverbs 9:10 LB).

God is the only living God today, therefore, He is the only true God.

When anyone sees the beauty and harmony of the Earth and Universe, he is observing God's handiwork. Creation speaks to us daily.

> The Heavens are telling the glory of God; they are a marvelous display of His craftsmanship. Day and night they keep on telling about God. Without a sound or word, silent in the skies, their message reaches out to all the world (Psalm 19:1-4 LB).

God's message is written on all of creation.

> For the truth about God is known to them instinctively; God has put this knowledge in their hearts. Since earliest times men have seen the earth and sky and all God made, and have known of His existence and great eternal power. So they will have no excuse when they stand before God at Judgment Day (Romans 1:19-20 LB).

God stated, in many places in the Bible, that He created the universe, Earth and all things. Abraham believed what God said to him and this was considered as righteousness; on account of His faith (Genesis 15:6). If one denies God as the Creator, then this unbelief would be considered as unrighteousness.

One must believe and accept God, who gave His son on the Cross for our sins and become a disciple and follower, in order to be saved. Church membership or special denominations are not as important as trusting and obedience. No Baptists, Methodists, Catholics, etc. will be in Heaven because of their denomination; only born again believers (John 3:3). No works or rituals are necessary since salvation is by grace through faith (Ephesians 2:8-9). We do, however, show our faith by our works. When one accepts the Lord with his whole heart, mind and soul (Matthew 22:37), he will experience His awesome power and be born again (becoming a new creation). The same force used in creation now fashions, energizes and changes one to a son or daughter of God (II Corinthians 6:18) by His cosmic redemption. When the power of the Spirit of God comes into your heart, you will never again doubt His existence. God will give you power to comprehend the error of the false, worldly and secular principles, such as Evolution, Big Bang or Old Earth theories. This spiritual power sheds a new light (the Light of the World) on the false, godless fabrications of the world. Satan would then no longer have control of a mind that was misled by intelligent, secular lies. The Gospel with all its beauty and simplicity is no longer hidden. The strongholds of evolutionary humanism, like the pagans' actions of old, will no longer prevail. A salvation experience occurs when one completely surrenders his life to the Lord and is filled with the Spirit of God. This experience will eliminate all doubt of God's existence and His eternal power. I have accepted Christ as my Savior and felt His power, love and grace. It is my greatest possession. He is ready to receive all who totally accept and believe in Him (John 3:16). One of the wisest men of all time, King Solomon, studied and

Conclusion

pondered life's great questions. His conclusion is written at the end of Ecclesiastes and I agree with his summation.

> But, my son, be warned: there is no end of opinions ready to be expressed. Studying them can go on forever, and become very exhausting! Here is my final conclusion: fear God and obey His commandments, for this is the entire duty of man. For God will judge us for everything we do, including every hidden thing, good or bad (Ecclesiastes 12:12-14 LB).

BIBLIOGRAPHY

Austin, Steven A. Ph.D., *Mt. St. Helens and Catastrophism*, Institute for Creation Research, Impact Article No. 157, July 1986.

Clanin, Gloria, *In the Days of Noah*, Master Books, 1999.

Foster, Robert J., *General Geology*, Fifth Edition, Merrill Publishing Company, 1988.

Freedman, David Noel, Editor-in-Chief, *Eerdmans Dictionary of the Bible*, Wm. B. Eerdmans Publishing Company, Grand Rapids, Michigan, 2000.

Gibbons, William J. Ph.D. and Dr. Kent Hovind, *Claws, Jaws and Dinosaurs.*

Gish, Duane T. Ph.D., *Evolution: The Fossils Still Say NO!*, 1995.

Hamblin, W. Kenneth, *The Earth's Dynamic Systems,* A Textbook in Physical Geology, Macmillan Publishing Company, 1989.

Morris, Dr. Henry M., *The Heavens Declare The Glory of God,* Daily Devotionals, World Publishing, Inc., 1997.

Morris, Dr. Henry M., *The DEFENDER'S Study Bible*, King James Version, World Publishing, Inc., 1995.

Morris, Dr. Henry M. and Dr. Martin E. Clark, *The Bible Has The Answer*, Master Books, 2001.

Morris, John D. Ph.D., *The Geology Book*, Master Books, 2000.

Peterson, Dennis R., *Unlocking the Mysteries of Creation*, Master Books, 1987.

USGS Earthquake Hazards Program: *Earthquake Facts and Statistics*, September 2, 2003, http://neic.usgs.gov/neis/eqlists/eqstats.html.

Walker, Bryce, *Planet Earth, Earthquake*, Time-Life Books, Inc., Alexandria, Virginia, 1984.

ILLUSTRATION INDEX

Table 4.1 Major Elements of the Earth's Crust... 39

Mohs' Scale of Hardness.............................. 40

Scale of Earthquake Intensity Chart 56

Richter Scale of Earthquake Magnitude........... 58

Total World Earthquakes Chart....................... 64

GLOSSARY

Age of Earth – This is calculated by creationists as being 6,000-10,000 years old and evolutionists as 4.6 billion years old.

Angels – A spiritual being created to serve God.

Basalt – A type of igneous rock that makes up most of the oceanic crust. On land it forms when extruded by volcanoes or through fissures.

Canopy – A frozen layer of ice that might have protected the Earth from cosmic rays and could have caused a greenhouse effect.

Catastrophism – The philosophy about the past, which allows for totally different processes and/or different process rates, scales, and intensities than those operating today. This includes the idea that processes such as creation and dynamic global flooding have shaped the entire planet.

Continental Drift – The concept that the continental plates have moved apart (or collided); for example, that Africa and South America were once connected.

Core – The center of the Earth is thought to be a sphere of iron and nickel, divided into two zones. The outer core is in molten or liquid form, while the inner core is solid.

Creation Day – A belief that a creation day was a 24-hour period.

Creationist – An individual that believes the God of the Bible created all things.

Crust – The thin covering of planet Earth, which includes the continents and ocean basins. Nowhere is it more than 60 miles (100 km) thick.

Earthquake – A sudden motion or trembling in the Earth.

Evolutionist – An individual that believes that life on Earth has developed gradually, over billions of years, from simple to complex forms.

Fault – A fracture in rock along which separation or movement has taken place.

Fold – A bend or flexure in a layer of rock.

Fossil – The direct or indirect remains of an animal or plant or any evidence of past life.

Geology – The science which studies the Earth, the rocks of which it is composed, and the changes which it has undergone or is undergoing.

Granite – A widespread igneous rock, which contains abundant quartz and feldspar, and makes up a significant portion of the continental crust.

Igneous Rock – Rock formed when hot, molten magma cools and solidifies.

Intelligent Design – The idea that our complex, beautiful world illustrates a created, well planned order of events.

Jehovah God – The God of the Bible described as the Creator of all things.

Magma – A naturally occurring mobile rock material, generated within the Earth.

Mantle – The zone beneath the thin crust and above the core of the earth. It is about 1,864 miles (3,000 km) thick.

Metamorphic Rock – These rocks are formed when heat, pressure, and/or chemical action alters previously existing rock.

Mineral – A naturally formed chemical element or compound having a definite chemical composition and crystal form.

Mount Ararat – A mountain in Turkey listed as the resting-place of Noah's Ark.

Mount St. Helens – A volcanic mountain located in Washington State USA. It was the scene of a violent volcanic destructive eruption in 1980.

Mount Sinai – An igneous rock mount, located between the horns of the Red Sea, on which Moses received the Ten Commandments.

Omnipotent – Having virtually unlimited authority, influence or power.

Omniscient – Possessed of universal or complete knowledge and wisdom.

Paleontology – The scientific study of fossils.

Plate – The Earth's crust, both continental and oceanic, is divided into plates, with boundaries identified by zones of earthquake activity. The idea of plate tectonics holds that these plates move relative to one another, sometimes separating or colliding, and sometimes moving past each other.

Radioisotope Dating – This is the attempt to determine a rock's age by measuring the ratio of radioactive isotopes and the rate at which they decay.

Rock – Any naturally formed aggregate or mass of mineral matter, constituting an essential part of the Earth's crust.

Satan – A fallen angel known as the leader of evil spirits.

Secular – This is related to the worldly or temporal; not openly or specifically religious.

Sedimentary Rock – Rock that is formed by the deposition and consolidation of loose particles of sediment, and those formed by precipitation from water. The best examples are Limestone, Sandstone and Shale.

Spontaneous Generation – The idea that with random process and time, the raw elements of the Earth become living organisms.

Stone – A piece of rock that is used for something.

Tsunami – Sometimes called a tidal wave. A seismic sea wave produced by an underwater disturbance such as an earthquake, volcano or landslide. They can be extremely destructive.

Uniformitarianism – The philosophy about the past which assumes no past events of a different nature than those possible today, and/or operating at rates, scales and intensities far greater than those operating today. The slogan "the present is the key to the past" characterizes this idea.